MyEngineeringLab™

Right now, in your course, there are young men and women whose engineering achievements could revolutionize, improve, and sustain future generations.

Don't Let Them Get Away.

Thinking Like an Engineer, Third Edition, together with MyEngineeringLab, is a complete solution for providing an engaging in-class experience that will inspire your students to stay in engineering, while also giving them the practice and scaffolding they need to keep up and be successful in the course.

Learn more at **www.myengineeringlab.com**

PEARSON ALWAYS LEARNING

Elizabeth A. Stephan • David R. Bowman • William J. Park
Benjamin L. Sill • Matthew W. Ohland

Thinking Like an Engineer

An Active Learning Approach

ENGR 1070, 1080

Fourth Custom Edition for Clemson University

Taken from:
Thinking Like an Engineer: An Active Learning Approach,
Third Edition
by Elizabeth A. Stephan, David R. Bowman, William J. Park,
Benjamin L. Sill, and Matthew W. Ohland

Cover Art: Blue guitar/Evgeny Guityaev/Shutterstock
X-ray of guitar/Gustoimages/Science Photo Library

Taken from:

Thinking Like an Engineer: An Active Learning Approach, Third Edition
by Elizabeth A. Stephan, David R. Bowman, William J. Park, Benjamin L. Sill, and Matthew W. Ohland

Upper Saddle River, New Jersey 07458

This special edition published in cooperation with Pearson Learning Solutions.

Pearson Learning Solutions, 501 Boylston Street, Suite 900, Boston, MA 02116
A Pearson Education Company
www.pearsoned.com

Printed in the United States of America

1 2 3 4 5 6 7 8 9 10 V092 16 15 14

000200010271907440

CN

ISBN 10: 1-269-91098-1
ISBN 13: 978-1-269-91098-9

NOTE TO STUDENT: The material in this book has been specifically selected by your instructors. They have chosen to include only material relevant to this course; and in so doing some references in the preface may not be applicable to this text.

CONTENTS

PREFACE vii

ACKNOWLEDGMENTS xv

Part 4
PUNCTILIOUS PROGRAMMING 525

SOME ADVANTAGES OF COMPUTERS 526

CHAPTER 15
ALGORITHMS 528

15.1 SCOPE 528

15.2 WRITTEN ALGORITHMS 530

15.3 GRAPHICAL ALGORITHMS 532

15.4 ALGORITHM BEST PRACTICES 537

IN-CLASS ACTIVITIES 544

REVIEW QUESTIONS 547

CHAPTER 16
MATLAB VARIABLES AND DATA TYPES 550

16.1 VARIABLE BASICS 551

16.2 NUMERIC TYPES AND SCALARS 553

16.3 VECTORS 557

16.4 MATRICES 566

16.5 CHARACTER STRINGS 574

16.6 CELL ARRAYS 577

16.7 STRUCTURE ARRAYS 584

16.8 SAVING AND RESTORING VALUES 587

IN-CLASS ACTIVITIES 589

REVIEW QUESTIONS 593

CHAPTER 17
PROGRAMS AND FUNCTIONS 596

17.1 PROGRAMS 596

17.2 FUNCTIONS 606

17.3 DEBUGGING MATLAB CODE 612

IN-CLASS ACTIVITIES 615

REVIEW QUESTIONS 621

CHAPTER 18
INPUT/OUTPUT IN MATLAB 627

18.1 INPUT 627

18.2 OUTPUT 633

18.3 PLOTTING 637

18.4 POLYFIT 644

18.5 MICROSOFT EXCEL I/O 650

IN-CLASS ACTIVITIES 655

REVIEW QUESTIONS 664

CHAPTER 19
LOGIC AND CONDITIONALS 673

19.1 RELATIONAL AND LOGICAL OPERATORS 674

19.2 LOGICAL VARIABLES 676

19.3 CONDITIONAL STATEMENTS IN MATLAB 682

19.4 `switch` STATEMENTS 686

19.5 ERRORS AND WARNINGS 689

IN-CLASS ACTIVITIES 692

REVIEW QUESTIONS 699

CHAPTER 20
LOOPING STRUCTURES 709

20.1 `for` LOOPS 709

20.2 `while` LOOPS 719

20.3 APPLICATION OF LOOPS: GUI 723

IN-CLASS ACTIVITIES 735

REVIEW QUESTIONS 744

COMPREHENSION CHECK ANSWERS 755

PREFACE

At our university, all students who wish to major in engineering begin in the General Engineering Program, and after completing a core set of classes, they can declare a specific engineering major. Within this core set of classes, students are required to take math, physics, chemistry, and a two-semester engineering sequence. Our courses have evolved to address not only the changing qualities of our students, but also the changing needs of our customers. The material taught in our courses is the foundation upon which the upper level courses depend for the skills necessary to master more advanced material. It was for these freshman courses that this text was created.

We didn't set out to write a textbook: we simply set out to find a better way to teach our students. Our philosophy was to help students move from a mode of learning, where everything was neatly presented as lecture and handouts where the instructor was looking for the "right" answer, to a mode of learning driven by self-guided inquiry. We wanted students to advance beyond "plug-and-chug" and memorization of problem-solving methods—to ask themselves if their approaches and answers make sense in the physical world. We couldn't settle on any textbooks we liked without patching materials together—one chapter from this text, four chapters from another—so we wrote our own notes. Through them, we tried to convey that engineering isn't always about having the answer—sometimes it's about asking the right questions, and we want students to learn how to ask those sorts of questions. Real-world problems rarely come with all of the information required for their solutions. Problems presented to engineers typically can't be solved by looking at how someone else solved the exact same problem. Part of the fun of engineering is that every problem presents a unique challenge and requires a unique solution. Engineering is also about arriving at an answer and being able to justify the "why" behind your choice, and equally important, the "why not" of the other choices.

We realized quickly, however, that some students are not able to learn without sufficient scaffolding. Structure and flexibility must be managed carefully. Too much structure results in rigidity and unnecessary uniformity of solutions. On the other hand, too much flexibility provides insufficient guidance, and students flounder down many blind alleys, thus making it more difficult to acquire new knowledge. The tension between these two must be managed constantly. We are a large public institution, and our student body is very diverse. Our hope is to provide each student with the amount of scaffolding they need to be successful. Some students will require more background work than others. Some students will need to work five problems, and others may need to work 50. We talk a great deal to our students about how each learner is unique. Some students need to listen to a lecture; some need to read the text over three times, and others just need to try a skill and make mistakes to discover what they still don't understand. We have tried to provide enough variety for each type of learner throughout.

Over the years, we have made difficult decisions on exactly what topics, and how much of each topic, to teach. We have refined our current text to focus on mastering four areas, each of which is introduced below.

PART 1: ENGINEERING ESSENTIALS

There are three threads that bind the first six chapters in Engineering Essentials together. The first is expressed in the part title: all are essential for a successful career in engineering. The second is communications. Part 1 concludes with an introduction to a problem-solving methodology.

First, as an aspiring engineer, it is important that students attempt to verify that engineering is not only a career that suits their abilities but also one in which they will find personal reward and satisfaction.

Second, practicing engineers often make decisions that will affect not only the lives of people but also the viability of the planetary ecosystem that affects all life on Earth. Without a firm grounding in making decisions based on ethical principles, there is an increased probability that undesirable or even disastrous consequences may occur.

Third, most engineering projects are too large for one individual to accomplish alone; thus, practicing engineers must learn to function effectively as a team, putting aside their personal differences and combining their unique talents, perspectives, and ideas to achieve the goal.

Finally, communications bind it all together. Communication, whether written, graphical, or spoken, is essential to success in engineering.

This part ends off where all good problem solving should begin—with estimation and a methodology. It's always best to have a good guess at any problem before trying to solve it more precisely. SOLVEM provides a framework for solving problems that encourages creative observation as well as methodological rigor.

PART 2: UBIQUITOUS UNITS

The world can be described using relatively few dimensions. We need to know what these are and how to use them to analyze engineering situations. Dimensions, however, are worthless in allowing engineers to find the numeric solution to a problem. Understanding units is essential to determine the correct numeric answers to problems. Different disciplines use different units to describe phenomena (particularly with respect to the properties of materials such as viscosity, thermal conductivity, density and so on). Engineers must know how to convert from one unit system to another. Knowledge of dimensions allows engineers to improve their problem-solving abilities by revealing the interplay of various parameters.

PART 3: SCRUPULOUS WORKSHEETS

When choosing an analysis tool to teach students, our first pick is Excel™. Students enter college with varying levels of experience with Excel. To allow students who are novice users to learn the basics without hindering more advanced users, we have placed the basics of Excel in the Appendix material, which is available online. To help students determine if they need to review the Appendix material, an activity has been included in the introductions to Chapter 10 (Worksheets), Chapter 11 (Graphing), and Chapter 12 (Trendlines) to direct students to Appendices B, C, and D, respectively.

Once students have mastered the basics, each chapter in this part provides a deeper usage of Excel in each category. Some of this material extends beyond a simple introduction to Excel, and often, we teach the material in this unit by jumping around, covering half of each chapter in the first semester, and the rest of the material in the second semester course.

Chapter 12 introduces students to the idea of similarities among the disciplines, and how understanding a theory in one application can often aid in understanding a similar theory in a different application. We also emphasize the understanding of models (trendlines) as possessing physical meaning. Chapter 13 discusses a process for determining a mathematical model when presented with experimental data and some advanced material on dealing with limitations of Excel.

Univariate statistics and statistical process control wrap up this part of the book by providing a way for engineering students to describe both distributions and trends.

PART 4: PUNCTILIOUS PROGRAMMING

Part 4 (Punctilious Programming) covers a variety of topics common to any introductory programming textbook. In contrast to a traditional programming textbook, this part approaches each topic from the perspective of how each can be used in unison with the others as a powerful engineering problem-solving tool. The topics presented in Part 4 are introduced as if the student has no prior programming ability and are continually reiterated throughout the remaining chapters.

For this textbook we chose MATLAB™ as the programming language because it is commonly used in many engineering curricula. The topics covered provide a solid foundation of how computers can be used as a tool for problem solving and provide enough scaffolding for transfer of programming knowledge into other languages commonly used by engineers (such as C/C++/Java).

THE "OTHER" STUFF WE'VE INCLUDED...

Throughout the book, we have included sections on surviving engineering, time management, goal setting, and study skills. We did not group them into a single chapter, but have scattered them throughout the part introductions to assist students on a topic when they are most likely to need it. For example, we find students are much more open to discuss time management in the middle of the semester rather than the beginning.

In addition, we have called upon many practicing and aspiring engineers to help us explain the "why" and "what" behind engineering. They offer their "Wise Words" throughout this text. We have included our own set of "Wise Words" as the introduction to each topic here as a glimpse of what inspired us to include certain topics.

NEW TO THIS EDITION

The third edition of *Thinking Like an Engineer: An Active Learning Approach* (TLAE) contains new material and revisions based off of the comments from faculty teaching with our textbook, the recommendations of the reviewers of our textbook, and most importantly, the feedback from our students. We continue to strive to include the latest software releases; in this edition, we have upgraded to Microsoft Office (Excel) 2013 and MATLAB 2013. We have added approximately 30% new questions. In addition, we have added new material that reflects the constant changing face of engineering education because many of our upperclassman teaching assistants frequently comment to us "I wish I had ___ when I took this class."

New to this edition, by chapter:

- Chapter 1: Everyday Engineering
  - New section on the field of Engineering Technology.
- Chapter 3: Design and Teamwork
  - New sequence of topics, to allow expanded discussion on defining the problem, determining criteria, brainstorming, making decisions and testing solutions.
- Chapter 8: Universal Units
  - New section on Electrical Concepts.
- Chapter 14: Statistics
  - Combined material from Chapters 14 (Excel) and 18 (MATLAB) in TLAE 2e to make a single unified chapter on Statistics.
- Chapter 16: Variables and Data Types
  - New material on the various ways MATLAB stores and processes data.
  - Selected material from TLAE 2e has been moved to this chapter, including cell arrays.
- Chapter 18: Input/Output in MATLAB
  - Combined material from Chapter 20 in TLAE 2e on using Microsoft Excel to input data to and output data from MATLAB.
- Chapter 19: Logic and Conditionals
  - New sections on Switch Statements and using Errors and Warnings.
- Online Appendix Materials
  - Umbrella Projects have all been moved online to allow for easier customizing of the project for each class.

HOW TO USE

As we have alluded to previously, this text contains many different types of instruction to address different types of learners. There are two main components to this text: hard copy and online.

In the hardcopy, the text is presented topically rather than sequentially, but hopefully with enough autonomy for each piece to stand alone. For example, we routinely discuss only part of the Excel material in our first semester course, and leave the rest to the second semester. We hope this will give you the flexibility to choose how deeply into any given topic you wish to dive, depending on the time you have, the starting abilities of your students, and the outcomes of your course. More information about topic sequence options can be found in the instructor's manual.

Within the text, there are several checkpoints for students to see if they understand the material. Within the reading are **Comprehension Checks**, with the answers provided in the back of the book. Our motivation for including Comprehension Checks within the text rather than include them as end of part questions is to maintain the active spirit of the classroom within the reading, allowing the students to self-evaluate their understanding of the material in preparation for class—to enable students to be self-directed learners, we must encourage them to self-evaluate regularly. At the end of each chapter, **In-Class Activities** are given to reinforce the material in each chapter. In-Class Activities exist to stimulate active conversation within pairs and groups of students working through the material. We generally keep the focus on student effort, and ask them to keep working the problem until they arrive at the right answer. This provides them with a set of worked out problems, using their own logic, before they are asked to tackle more difficult problems. The **Review** sections provide additional questions, often combining skills in the current chapter with previous concepts to help students climb to the next level of understanding. By providing these three types of practice, students are encouraged to reflect on their understanding in preparing for class, during class, and at the end of each chapter as they prepare to transfer their knowledge to other areas. Finally we have provided a series of **Umbrella Projects** to allow students to apply skills that they have mastered to larger-scope problems. We have found the use of these problems extremely helpful in providing context for the skills that they learn throughout a unit.

Understanding that every student learns differently, we have included several media components in addition to traditional text. Each section within each chapter has an accompanying set of **video lecture slides**. Within these slides, the examples presented are unique from those in the text to provide another set of sample solutions. The slides are presented with **voiceover**, which has allowed us to move away from traditional in-class lecture. We expect the students to listen to the slides outside of class, and then in class we typically spend time working problems, reviewing assigned problems, and providing **"wrap-up" lectures**, which are mini-versions of the full lectures to summarize what they should have gotten from the assignment. We expect the students to come to class with questions from the reading and lecture that we can then help clarify. We find with this method, the students pay more attention, as the terms and problems are already familiar to them, and they are more able to verbalize what they don't know. Furthermore, they can always go back and listen to the lectures again to reinforce their knowledge as many times as they need.

Some sections of this text are difficult to lecture, and students will learn this material best by **working through examples**. This is especially true with Excel and MATLAB, so you will notice that many of the lectures in these sections are shorter than previous material. The examples are scripted the first time a skill is presented, and students are expected to have their laptop open and work through the examples (not just read them). When students ask us questions in this section, we often start the answer by asking them to "show us your work from Chapter XX." If the student has not actually worked the examples in that chapter, we tell them to do so first; often, this will answer their questions.

After the first few basic problems, in many cases where we are discussing more advanced skills than data entry, we have **provided starting worksheets and code** in the online version by "hanging" the worksheets within the online text. Students can access the starting data through the online copy of the book. In some cases, though, it is difficult to explain a skill on paper, or even with slides, so for these instances we have included **videos**.

Finally, for the communication section, we have provided **templates** for several types of reports and presentations. These can also be accessed in the Pearson eText version, available with adoption of MyEngineeringLab™. Visit www.pearsonhighered.com/TLAE for more information.

MyEngineeringLab™

Thinking Like an Engineer, Third Edition, together with MyEngineeringLab provides an engaging in-class experience that will inspire your students to stay in engineering, while also giving them the practice and scaffolding they need to keep up and be successful in the course. It's a complete digital solution featuring:

- A customized study plan for each student with remediation activities provides an opportunity for self paced learning for students at all different levels of preparedness.
- Automatically graded homework review problems from the book and self study quizzes give immediate feedback to the student and provide comprehensive gradebook tracking for instructors.
- Interactive tutorials with additional algorithmically generated exercises provide opportunity for point-of-use help and for more practice.
- "Show My Work" feature allows instructors to see the entire solution, not only the graded answer.
- Learning objectives mapped to ABET outcomes provide comprehensive reporting tools.
- Selected spreadsheet exercises are provided in a simulated Excel environment; these exercises are automatically graded and reported back to the gradebook.
- Pre-built writing assignments provide a single place to create, track, and grade writing assignments, provide writing resources, and exchange meaningful, personalized feedback to students.
- Available with or without the full eText.

If adopted, access to MyEngineeringLab can be bundled with the book or purchased separately. For a fully digital offering, learn more at www.myengineeringlab.com or www.pearsonhighered.com/TLAE.

ADDITIONAL RESOURCES FOR INSTRUCTORS

Instructor's Manual—Available to all adopters, this provides a complete set of solutions for all activities and review exercises. For the In-Class Activities, suggested guided inquiry questions along with time frame guidelines are included. Suggested content sequencing and descriptions of how to couple assignments to the Umbrella Projects are also provided.

PowerPoints—A complete set of lecture PowerPoint slides make course planning as easy as possible.

Sample Exams—Available to all adopters, these will assist in creating tests and quizzes for student assessment.

MyEngineeringLab—Provides web-based assessment, tutorial, homework and course management. www.myengineeringlab.com

All requests for instructor resources are verified against our customer database and/or through contacting the requestor's institution. Contact your local Pearson/Prentice Hall representative for additional information.

WHAT DOES THINKING LIKE AN ENGINEER MEAN?

We are often asked about the title of the book. We thought we'd take a minute and explain what this means, to each of us. Our responses are included in alphabetical order.

> *For me, thinking like an engineer is about creatively finding a solution to some problem. In my pre-college days, I was very excited about music. I began my musical pursuits by learning the fundamentals of music theory by playing in middle school band and eventually worked my way into different bands in high school (orchestra, marching and, jazz band; and branching off into teaching myself how to play guitar. I love playing and listening to music because it gives me an outlet to create and discover art. I pursued engineering for the same reason; as an engineer, you work in a field that creates or improves designs or processes. For me, thinking like an engineer is exactly like thinking like a musician—through my fundamentals, I'm able to be creative, yet methodical, in my solutions to problems.*
>
> D. Bowman, Computer Engineer

> *Thinking like an engineer is about solving problems with whatever resources are most available—or fixing something that has broken with materials that are just lying around. Sometimes, it's about thinking ahead and realizing what's going to happen before something breaks or someone gets hurt—particularly in thinking about what it means to fail safe—to design how something will fail when it fails. Thinking like an engineer is figuring out how to communicate technical issues in a way that anyone can understand. It's about developing an instinct to protect the public trust—an integrity that emerges automatically.*
>
> M. Ohland, Civil Engineer

> *To me, understanding the way things work is the foundation on which all engineering is based. Although most engineers focus on technical topics related to their specific discipline, this understanding is not restricted to any specific field, but applies to everything! One never knows when some seemingly random bit of knowledge, or some pattern discerned in a completely disparate field of inquiry, may prove critical in solving an engineering problem. Whether the field of investigation is Fourier analysis, orbital mechanics, Hebert boxes, personality types, the Chinese language, the life cycle of mycetozoans, or the evolution of the music of Western civilization, the more you understand about things, the more effective an engineer you can be. Thus, for me, thinking like an engineer is intimately, inextricably, and inexorably intertwined with the Quest for Knowledge. Besides, the world is a truly fascinating place if one bothers to take the time to investigate it.*
>
> W. Park, Electrical Engineer

Engineering is a bit like the game of golf. No two shots are ever exactly the same. In engineering, no two problems or designs are ever exactly the same. To be successful, engineers need a bag of clubs (math, chemistry, physics, English, social studies) and then need to have the training to be able to select the right combination of clubs to move from the tee to the green and make a par (or if we are lucky, a birdie). In short, engineers need to be taught to THINK.

B. Sill, Aerospace Engineer

I like to refer to engineering as the color grey. Many students enter engineering because they are "good at math and science." I like to refer to these disciplines as black and white—there is one way to integrate an equation and one way to balance a chemical reaction. Engineering is grey, a blend of math and science that does not necessarily have one clear answer. The answer can change depending on the criteria of the problem. Thinking like an engineer is about training your mind to conduct the methodical process of problem solving. It is examining a problem from many different angles, considering the good, the bad and the ugly in every process or product. It is thinking creatively to discover ways of solving problems, or preventing issues from becoming problems. It's about finding a solution in the grey and presenting it in black and white.

E. Stephan, Chemical Engineer

Lead author note: When writing this preface, I asked each of my co-authors to answer this question. As usual, I got a wide variety of interpretations and answers. This is typical of the way we approach everything we do, except that I usually try and mesh the responses into one voice. In this instance, I let each response remain unique. As you progress throughout this text, you will (hopefully) see glimpses of each of us interwoven with the one voice. We hope that through our uniqueness, we can each reach a different group of students and present a balanced approach to problem solving, and, hopefully, every student can identify with at least one of us.

—Beth Stephan
Clemson University
Clemson, SC

ACKNOWLEDGMENTS

When we set out to formalize our instructional work, we wanted to portray engineering as a reality, not the typical flashy fantasy portrayed by most media forums. We called on many of our professional and personal relationships to help us present engineering in everyday terms. During a lecture to our freshman, Dr. Ed Sutt [PopSci's 2006 Inventor of the Year for the HurriQuake Nail] gave the following advice: ***A good engineer can reach an answer in two calls: the first, to find out who the expert is; the second, to talk to the expert.*** Realizing we are not experts, we have called on many folks to contribute articles. To our experts who contributed articles for this text, we thank: Dr. Lisa Benson, Dr. Neil Burton, Jan Comfort, Jessica (Pelfrey) Creel, Jason Huggins, Leidy Klotz, and Troy Nunmaker.

To Dr. Lisa Benson, thank you for allowing us to use "Science as Art" for the basis of many photos that we have chosen for this text. To explain "Science as Art": *Sometimes, science and art meet in the middle. For example, when a visual representation of science or technology has an unexpected aesthetic appeal, it becomes a connection for scientists, artists and the general public. In celebration of this connection, Clemson University faculty and students are challenged to share powerful and inspiring visual images produced in laboratories and workspaces for the "Science as Art" exhibit.* For more information, please visit www.scienceasart.org. To the creators of the art, thank you for letting us showcase your work in this text: Martin Beagley, Dr. Caye Drapcho, Eric Fenimore, Dr. Scott Husson, Dr. Jaishankar Kutty, Dr. Kathleen Richardson, and Dr. Ken Webb. A special thanks Russ Werneth for getting us the great Hubble teamwork photo.

To the Rutland Institute for Ethics at Clemson University: The four-step procedure outlined in Chapter 2 on Ethics is based on the toolbox approach presented in the Ethics Across the Curriculum Seminar. Our thanks to Dr. Daniel Wueste, Director, and the other Rutlanders (Kelly Smith, Stephen Satris and Charlie Starkey) for their input into this chapter.

To Jonathan Feinberg and all the contributors to the Wordle (http://www.wordle.net) project, thank you for the tools to create for the Wordle images in the introduction sections. We hope our readers enjoy this unique way of presenting information, and are inspired to create their own Wordle!

To our friends and former students who contributed their Wise Words: Tyler Andrews, Corey Balon, Ed Basta, Sergey Belous, Brittany Brubaker, Tim Burns, Ashley Childers, Jeremy Comardelle, Matt Cuica, Jeff Dabling, Christina Darling, Ed D'Avignon, Brian Dieringer, Lauren Edwards, Andrew Flowerday, Stacey Forkner, Victor Gallas Cervo, Lisa Gascoigne, Khadijah Glast, Tad Hardy, Colleen Hill, Tom Hill, Becky Holcomb, Beth Holloway, Selden Houghton, Allison Hu, Ryan Izard, Lindy Johnson, Darryl Jones, Maria Koon, Rob Kriener, Jim Kronberg, Rachel Lanoie, Mai Lauer, Jack Meena, Alan Passman, Mike Peterson, Candace Pringle, Derek Rollend,

Eric Roper, Jake Sadie, Janna Sandel, Ellen Styles, Adam Thompson, Kaycie (Smith) Timmons, Devin Walford, Russ Werneth, and Aynsley Zollinger.

To our fellow faculty members, for providing inspiration, ideas, and helping us find countless mistakes: Dr. Steve Brandon, Dr. Ashley Childers, Andrew Clarke, Dr. David Ewing, Dr. Sarah Grigg, Dr. Richard Groff, Dr. Apoorva Kapadia, Dr. Sabrina Lau, Dr. Jonathan Maier, Dr. William Martin, Jessica Merino, and John Minor. You guys are the other half of this team that makes this the best place on earth to work! We could not have done this without you.

To the staff of the GE program, we thank you for your support of us and our students: Kelli Blankenship, Lib Crockett, Chris Porter, and all of our terrific advising staff both past and present. To the administration at Clemson, we thank you for your continued support of our program: Associate Dean Dr. Randy Collins, Interim Director Dr. Don Beasley, Dean Dr. Anand Gramopadhye, Provost Nadim Aziz. Special thanks to President Jim Barker for his inspirational leadership of staying the course and giving meaning to "One Clemson." We wish him all the best as he retired from the Presidency this December.

To the thousands of students who used this text in various forms over the years—thanks for your patience, your suggestions, and your criticism. You have each contributed not only to the book, but to our personal inspirations to keep doing what we do.

To all the reviewers who provided such valuable feedback to help us improve. We appreciate the time and energy needed to review this material, and your thoughtful comments have helped push us to become better.

To the great folks at Prentice Hall—this project would not be a reality without all your hard work. To Eric Hakanson, without that chance meeting this project would not have begun! Thanks to Holly Stark for her belief in this project and in us! Thanks to Scott Disanno for keeping us on track and having such a great vision to display our hard work. You have put in countless hours on this edition—thanks for making us look great! Thanks to Tim Galligan and the fabulous Pearson sales team all over the country for promoting our book to other schools and helping us allow so many students to start "Thinking Like Engineers"! We would not have made it through this without all of the Pearson team efforts and encouragement!

FINALLY, ON A PERSONAL NOTE

DRB: Thanks to my parents and sister for supporting my creative endeavors with nothing but encouragement and enthusiasm. To my grandparents, who value science, engineering, and education to be the most important fields of study. To my co-authors, who continue to teach me to think like an engineer. To Dana, you are the glue that keeps me from falling to pieces. Thank you for your support, love, laughter, inspiration, and determination, among many other things. You are entirely too rad. I love you.

MWO: My wife Emily has my love, admiration, and gratitude for all she does, including holding the family together. For my children, who share me with my students—Charlotte, whose "old soul" touches all who take the time to know her; Carson, who is quietly inspiring; and Anders, whose love of life and people endears him to all. I acknowledge my father Theodor, who inspired me to be an educator; my mother Nancy, who helped me understand people; my sister Karen, who lit a pathway in engineering; my brother Erik, who showed me that one doesn't need to be loud to be a leader; and my mother-in-law Nancy Winfrey, who shared the wisdom of a long career. I recognize those who helped me create an engineering education career path: Fred Orthlieb, Civil and Coastal

Engineering at the University of Florida, Marc Hoit, Duane Ellifritt, Cliff Hays, Mary Grace Kantowski, and John Lybas, the NSF's SUCCEED Coalition, Tim Anderson, Clemson's College of Engineering and Science and General Engineering, Steve Melsheimer, Ben Sill, and Purdue's School of Engineering Education.

WJP: Choosing only a few folks to include in an acknowledgment is a seriously difficult task, but I have managed to reduce it to five. First, Beth Stephan has been the guiding force behind this project, without whom it would never have come to fruition. In addition, she has shown amazing patience in putting up with my shenanigans and my weird perspectives. Next, although we exist in totally different realities, my parents have always supported me, particularly when I was a newly married, destitute graduate student fresh off the farm. Third, my son Isaac, who has the admirable quality of being willing to confront me with the truth when I am behaving badly, and for this I am grateful. Finally, and certainly most importantly, to Lila, my partner of more than one-third century, I owe a debt beyond anything I could put into words. Although life with her has seldom been easy, her influence has made me a dramatically better person.

BLS: To my amazing family, who always picked up the slack when I was off doing "creative" things, goes all my gratitude. To Anna and Allison, you are wonderful daughters who both endured and "experienced" the development of many "in class, hands on" activities—know that I love you and thank you. To Lois who has always been there with her support and without whining for over 40 years, all my love. Finally, to my co-authors who have tolerated my eccentricities and occasional tardiness with only minimum grumbling, you make great teammates.

EAS: To my co-authors, for tolerating all my strange demands, my sleep-deprived ravings and the occasional "I need this now" hysteria—and it has gotten worse with the third edition—you guys are the best! To my mom, Kay and Denny—thanks for your love and support. To Khadijah & Steven, wishes for you to continue to conquer the world! To Brock and Katie, I love you both a bushel and a peck. You are the best kids in the world, and the older you get the more you inspire me to be great at my job. Thank you for putting up with all the late nights, the lack of home-cooked meals, and the mature-beyond-your-years requirements I've asked of you. Finally, to Sean . . . last time I swore the rough parts were done, but man this edition was tough to finish up! I love you more than I can say—and know that even when I forget to say it, I still believe in us. "Show a little faith, there's magic in the night . . ."

Part 4 PUNCTILIOUS PROGRAMMING

Chapter 15
ALGORITHMS
15.1 SCOPE
15.2 WRITTEN ALGORITHMS
15.3 GRAPHICAL ALGORITHMS
15.4 ALGORITHM BEST PRACTICES

Chapter 16
MATLAB VARIABLES AND DATA TYPES
16.1 VARIABLE BASICS
16.2 NUMERIC TYPES AND SCALARS
16.3 VECTORS
16.4 MATRICES
16.5 CHARACTER STRINGS
16.6 CELL ARRAYS
16.7 STRUCTURE ARRAYS
16.8 SAVING AND RESTORING VALUES

Chapter 17
PROGRAMS AND FUNCTIONS
17.1 PROGRAMS
17.2 FUNCTIONS
17.3 DEBUGGING MATLAB CODE

Chapter 18
INPUT/OUTPUT IN MATLAB
18.1 INPUT
18.2 OUTPUT
18.3 PLOTTING
18.4 POLYFIT
18.5 MICROSOFT EXCEL I/O

Chapter 19
LOGIC AND CONDITIONALS
19.1 RELATIONAL AND LOGICAL OPERATORS
19.2 LOGICAL VARIABLES
19.3 CONDITIONAL STATEMENTS IN MATLAB
19.4 `SWITCH` STATEMENTS
19.5 ERRORS AND WARNINGS

LEARNING OBJECTIVES

The overall learning objectives for this part include:

Chapter 15:

- Defining the scope of a problem and creating a written or graphical algorithm to solve the problem.

Chapter 16:

- Understand the various methods of storing information in MATLAB.
- Performing basic matrix operations.

Chapter 17:

- Writing MATLAB programs and / or functions to solve engineering problems.
- Reading and interpreting MATLAB programs written by others.
- Debugging a program to identify different types of errors.

Chapter 18:

- Writing input statements to allow the user to interact with the MATLAB environment.
- Write output statements to inform the user of program outcomes.
- Create graphs and use trendlines to enhance problem solving.
- Read data and record results between MATLAB and Microsoft Excel environments.

Chapter 19:

- Use conditional statements and switch statements to automate decision making.
- Use error and warning statements to aid the user in program execution.

Chapter 20:

- Use looping structures (for and while) to write eliminate large blocks of repetitive code.
- Use a GUI to aid the user in interacting with the MATLAB environment.

Chapter 20
LOOPING STRUCTURES
20.1 FOR LOOPS
20.2 WHILE LOOPS
20.3 APPLICATION OF LOOPS: GUI

Computers are controlled by software that can be designed in a variety of programming languages. Computer programs are a translation of what you want to accomplish into something the computer can understand, so the term "programming language" is particularly appropriate. Some computer programs are installed permanently or temporarily on computer chips, and others are installed on a variety of other media, such as hard drives or removable media like CD-ROMs.

Computers relentlessly produce a particular result given a particular set of input conditions. It can be frustrating when you make a simple mistake in a computer program—the computer will do exactly what you tell it to do, even if your mistake would be obvious to a person.

The biggest difference between a computer and a person is that you can ask a person open-ended questions—questions like design questions that can have many answers. Computers can only process questions that have a single answer.

This makes the process of programming a computer a bit like trying to ask another person to solve a problem when they are on the other side of a wall and you can communicate only by passing them slips of paper asking questions that can have only one answer and waiting for the person to pass back a slip of paper with the answer on it.

If a computer always produces the same result every time given the same input conditions, then why does my computer crash sometimes when I am doing something that should work?

The computers you use are simultaneously running a large number of complicated computer programs, including the operating system, background programs, and whatever programs you have started intentionally. Sometimes these programs compete for resources, causing a conflict. Other times, programs are complicated enough that the "input conditions," including the configuration of data in memory and on the hard disk, the time on the system clock, and other factors that change all the time while the computer is running create a combination of circumstances that the programmers never anticipated and so did not include programming code to handle, and the system crashes.

SOME ADVANTAGES OF COMPUTERS

Given our description of how computers work, it may sound to some as if computers are too simple to be useful. The value of programming is linked to a few important characteristics of computers.

- ***Calculation speed:*** Although computers can only answer analytical questions, they can answer such questions very quickly—in small fractions of a second. Computer programs can therefore ask the computer a lot of questions in a short time, and thus find the answer to more complicated problems by breaking down the complicated question into a series of simple questions.
- ***Information storage:*** In "Memory: Science Achieves Important New Insights into the Mother of the Muses" (*Newsweek*, September 29, 1986), Sharon Begley estimates that the mind can store an estimated 100 trillion bits of information. The typical computer has a small amount of storage compared to that, but computers are gaining. Where computers have a bigger advantage is that new information can be incorporated in a fraction of the time it takes a human to learn it.
- ***Information recall:*** Computers have nearly 100% recall of information, limited only by media failures. The human brain can be challenged to recall information in exactly the same form as it was stored.

WISE WORDS: WOULD YOU CONSIDER YOUR CURRENT POSITION TO BE "PURE ENGINEERING," A "BLEND OF ENGINEERING AND ANOTHER FIELD," OR "ANOTHER FIELD?"

I always feel my work is not "pure engineering," but rather often a blend of engineering, sales, accounting, research, inspection, and maintenance.

E. Basta, Material Engineer

I would consider my career in another field from engineering, however, highly reliant on my engineering background. As a management consultant, I have to break down complex problems, develop hypotheses, collect data I believe will prove or disprove the hypotheses, and perform the analysis. My focus area is companies who develop highly engineered products.

M. Ciuca, ME

My position is mostly pure engineering.

E. D'Avignon, CpE

I work in a blend of engineering and business. I spend most of my time working on business-related activities—forecasting, variance reporting, and timing/work decisions—but I also have to work closely with our field engineers and understand our project scopes. I use both my business and engineering knowledge on a daily basis—without each, I would not be able to succeed at my job.

R. Holcomb, IE

It is definitely a blend of engineering and law with a heavy dose of technical writing. It takes the thinking of an engineer or scientist to truly comprehend the inventions and the skill of a writer to convey the inventor's ideas in written and image terms that others will understand (including juries of lay people). It takes the thinking of a lawyer to come up with creative strategies and solutions when faced with a certain set of facts.

M. Lauer, EnvE

My current position is definitely a blend of engineering and at least one other field, but more like five other fields. I definitely use my engineering background in the way I think, the way I analyze data, how I approach problems, and how I integrate seemingly unrelated information together. The project management skills that I learned in engineering are helpful, too.

B. Holloway, ME

Even though my boss calls Hydrology "Voodoo Engineering," it is pure engineering.

J. Meena, CE

A blend of mechanical/aerospace engineering and human factors engineering—and management.

R. Werneth, ME

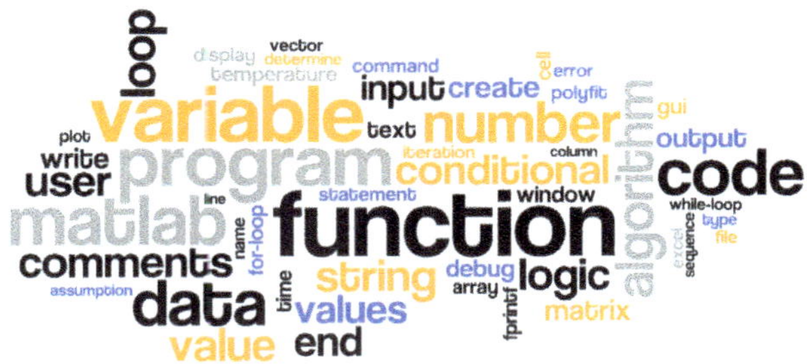

CHAPTER 15
ALGORITHMS

Learning to create effective algorithms is a crucial skill for any aspiring engineer. In general, an **algorithm** is a well-defined sequence of instructions that describe a process. Algorithms can be observed in everyday life through oral directions ("Simon says: raise your right hand"), written recipes ("Bake for 15 minutes at 350 degrees Fahrenheit"), graphical assembly instructions, or other graphical cues. As an engineer, writing any algorithm requires a complete understanding of all the necessary actions and decisions that must occur to complete a task.

When writing an algorithm, you must answer a few questions before attempting to design the process. To even begin thinking of a strategy to describe a process, you must have carefully defined the scope of the problem. The **scope** of an algorithm is the overall perspective and result that the algorithm must include in its design.

For example, if we are required to "sum all numbers between 1 and 5," before thinking about an approach to solve the problem, we must first determine if the scope is properly defined. Does the word "between" imply that 1 and 5 are included in the sum? Do "numbers" include only the integer values? What about the irrational numerical values? Clearly, we observe that we cannot properly define the scope of the charge to add all numbers between 1 and 5.

Likewise, imagine you were charged to design a device that transports people from Atlanta, Georgia, to Los Angeles, California. How many people must the device transport? Does the device need to travel on land? Should it travel by air? Should it travel by water? Does the device require any human interaction?

This section covers two methods of defining a process: with **written algorithms** and with **graphical algorithms**. Both methods require properly identifying the scope of the problem and all of the necessary input and output of the process.

15.1 SCOPE

LEARN TO:
- Define the scope of a problem
- Define known and unknown quantities in a problem
- Document any assumptions necessary to solve a problem

One of the most difficult steps in designing an algorithm is properly identifying the entire scope of the solution. Like solving a problem on paper involving unit conversions and equations, it is often necessary to state all of the known and unknown variables in order to determine a smart solution to the problem. If information is left out of the problem, it might be necessary to state an assumption in order to proceed with a solution. After all variables and assumptions about the problem have been identified, it is then possible to create a sequence of actions and decisions to solve the problem.

To clearly understand the scope of the problem, we often find it helpful to formally write out the known and unknown information, as well as state any assumptions necessary to solve the problem. In the following examples, notice that as the problem statements become more and more refined, the number of necessary assumptions decreases and eventually disappears.

EXAMPLE 15-1

For the problem statement, list all knowns, unknowns, and assumptions.

Problem: Sum all numbers between 1 and 10.

Known:

- *The minimum value in the sum will be 1.*
- *The maximum value in the sum will be 10.*

Unknown:

- *The sum of the sequence of numbers.*

Assumptions:

- *We will only include the whole number values (e.g., 1, 2, 3, . . .) in the sum.*
- *The sum will include the starting value of 1 and the ending value of 10.*

EXAMPLE 15-2

For the problem statement, list all knowns, unknowns, and assumptions.

Problem: Sum all numbers between (and including) 1 and 10.

Known:

- *The minimum value in the sum will be 1.*
- *The maximum value in the sum will be 10.*

Unknown:

- *The sum of the sequence of numbers.*

Assumptions:

- *We will only include the whole number values (e.g., 1, 2, 3, . . .) in the sum.*

EXAMPLE 15-3

For the problem statement, list all knowns, unknowns, and assumptions.

Problem: Sum all whole numbers between (and including) 1 and 10.

Known:

- *The minimum value in the sum will be 1.*
- *The maximum value in the sum will be 10.*

Unknown:

- *The sum of the sequence of numbers.*

Assumptions:

- *[None]*

COMPREHENSION CHECK 15-1

For the problem statement, list all knowns, unknowns, and assumptions. Problem: Sum all even numbers between (and including) 2 and 20.

COMPREHENSION CHECK 15-2

For the problem statement, list all knowns, unknowns, and assumptions. Problem: Multiply all powers of 5 between (and including) 5 and 50.

15.2 WRITTEN ALGORITHMS

LEARN TO:
- Create a linear written algorithm to solve an engineering problem
- Create a written algorithm that implements decision pathways
- Define the terms feedback loop and indefinite feedback loop

A written algorithm is a narrative set of instructions required to solve a problem. In everyday life, we encounter written algorithms in the form of oral instructions or written recipes. However, it is extremely common for humans to "fill in the blanks" on a poorly written algorithm. Imagine you are handed a strongly guarded family recipe for tacos. One of the steps in the archaic recipe is to "cook beef on low heat until done." To the veteran cook, it is apparent that this step requires cooking the prepared ground beef on a stovetop in a sauce pan for approximately 10 minutes on a burner setting of 2 to 3. To a first-time cook, the step is poorly defined and could result in potentially inedible taco meat.

Engineers and Written Algorithms

As an engineer, to write effective algorithms you must ensure that every step you include in a written algorithm must not be subject to misinterpretation. It is helpful to write an algorithm as if it were to be read by someone completely unfamiliar with the topic. Each step in the written algorithm should be written such that the stepwise scope is properly defined. The **stepwise scope** is all of the known and unknown information at that point in the procedure. If a step in an algorithm contains an assumption, you must formally declare it before proceeding with the next step. By ensuring that the stepwise scope is well defined, you ensure that your algorithm will not be subject to misinterpretation.

All written algorithms should be expressed sequentially. The most effective algorithms are written with many ordered steps, wherein each step contains one piece of information or procedure. While the author of an algorithm may consider each step in an algorithm to be "simple," it might not be trivial to an external interpreter. When writing an algorithm, it is helpful to assume that the reader of your algorithm can only perform small, simple tasks. Assume that your algorithm can be interpreted by a computer. A computer can execute small tasks efficiently and quickly, but unlike a human, a computer cannot fill in the blanks with information you intended the reader to assume.

Decision-making can be expressed in a written algorithm. Assume you are designing a process to determine if the value read from a temperature sensor in a vehicle indicates it is unsafe for operation. To express the decision in a written algorithm, phrase your decisions in questions that have a "Yes" or "No" response.

Format of Written Algorithms

The first step in writing any algorithm is defining the scope of the problem. After you define the scope of the problem, create an *ordered* or *bulleted list* of actions and decisions. Imagine taking an English class and writing a research report on the influence of 19th-century writers on modern-day fiction authors. Before writing the paper, you would create an outline to ensure that your topics have connectivity and flow. Just like the outline of an English paper, an algorithm is best expressed as a sequential list rather than as complete paragraphs of information.

If a decision is required in the algorithm, indent the actions to indicate the action is only associated with the particular condition.

EXAMPLE 15-4

Create a written algorithm to express a temperature given in relative units [°F or °C] in the corresponding absolute units [K or °R].

Known:

- *Temperature in relative units (degrees Celsius or degrees Fahrenheit).*

Unknown:

- *Temperature in absolute units (kelvins or degrees Rankine).*

Assumptions:

- *Since the problem does not explicitly state the temperature of interest, assume that the interpreter of the algorithm will input the temperature and units.*

Algorithm:

1. *Input the numeric value of the temperature.*
2. *Input the units of the numeric value of the temperature.*
3. *Ask if the input unit is degrees Fahrenheit.*
   (a) *If yes, calculate the value in degrees Rankine.*
   (b) *If no, calculate the value in kelvins.*
4. *Display the new value and absolute unit.*
5. *End the process.*

EXAMPLE 15-5

Create a written algorithm to calculate the sum of a sequence of whole numbers, given the upper and lower bounds of the sequence.

Known:

- *Upper bound of whole number sequence.*
- *Lower bound of whole number sequence.*

Unknown:

- *Sum of all whole numbers between the upper and lower bound.*

Assumptions:

- *Since the problem does not explicitly state the upper and lower bounds, assume the interpreter will ask for the values.*
- *Include the boundary values in the summation.*

Algorithm:

1. *Input the lower bound of the sequence.*
2. *Input the upper bound of the sequence.*
3. *If the lower bound is larger than the upper bound,*
   (a) *Warn the user that the input is invalid.*
   (b) *End the process.*
4. *If the upper bound is larger than the lower bound,*
   (a) *Create a variable to keep track of the sum (S).*
   (b) *Create a variable to keep track of the location in the sequence (L).*
   (c) *Set the initial value of S to be zero.*
   (d) *Set the initial value of L to be the lower bound.*
   (e) *If the value of L is less than or equal to the upper bound,*
      (i) *Add L to the current value of S.*
      (ii) *Add one to the current value of L.*
      (iii) *Return to step 4.e. and ask the question again.*
   (f) *If the value of L is greater than the upper bound,*
      (i) *Display the sum of the sequence (S).*
5. *End the process.*

In step 4.e.iii, we required that the interpreter return to an earlier step in the algorithm after changing the values of our variables. This allows us to create a **feedback loop** necessary to calculate the sequence of values. A feedback loop is a return to an earlier location in an algorithm with updated values of variables. It is important to note that if we failed to update the variables, the feedback loop will never terminate. A nonterminating feedback loop is also known as an **infinite feedback loop**.

COMPREHENSION CHECK 15-3

Create a written algorithm to multiply all integer powers of 5, 5^x, for x between (and including) 5 and 50.

15.3 GRAPHICAL ALGORITHMS

LEARN TO:
- Sketch a flowchart that implements a linear algorithm
- Sketch a flowchart that implements decision pathways
- Recognize and interpret shapes used in a graphical algorithm

To visualize a process, a graphical representation of algorithms is used instead of a written algorithm. A **flowchart** is a graphical representation of a written algorithm that describes the sequence of actions, decisions, and path of a process. Designing a flowchart forces the author of the algorithm to create small steps that can be quickly evaluated by the interpreter and enforces a sequence of all actions and decisions. Flowcharts are used by many different disciplines of engineering to describe different types of processes, so learning to create and interpret flowcharts is a critical skill for a young engineer. In fact, in the United States, any engineers who discover a new innovative algorithm can submit their concept to a patent office by representing the process in terms of a flowchart.

Three different shapes are used in the creation of flowcharts in this book; a number of other widely used operators are encountered across the world. In this book, we describe all actions with rectangles, all decisions with diamonds, and all connections between shapes with directional arrows.

Rules for Creating a Proper Flowchart

- The flowchart must contain a START rectangle to designate the beginning of a process.
- All actions must be contained within rectangles.
- All decisions must be contained within diamonds.
- All shapes must be connected by a one-way directional arrow.
- The flowchart must contain an END rectangle to designate the end of a process.

Actions

Actions are any executable steps in an algorithm that do not require a decision to be made. Based on this definition, any defined variables, calculations, and input or output commands would all be contained within action rectangles.

All simple actions are contained within a single rectangle on the flowchart. For each rectangle, two arrows are always associated with the shape, with two exceptions. The inward arrow to the rectangle represents the input to the action. It is assumed that any variables defined in the stepwise scope of an action rectangle are accessible and can be used in the action. The outward arrow from the rectangle represents the output of the action. If any new variables or calculations are performed within the rectangle, those values are passed along to the next shape's stepwise scope.

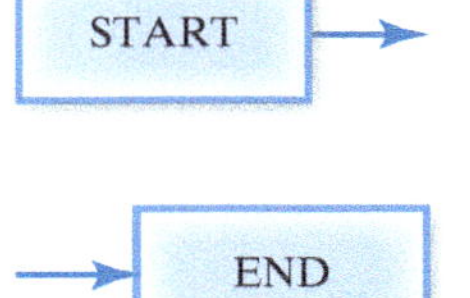

- Exception One: The START rectangle represents the beginning of the flowchart and does not contain an inward arrow. An oval shape is also commonly used to represent the start of an algorithm.
- Exception Two: The END rectangle represents the end of the flowchart and does not contain an outward output arrow. An oval shape is also commonly used to represent the end of an algorithm.

Decisions

Decisions are any executable steps in an algorithm that require the answer to a question with "Yes" or "No." All decisions in a flowchart must be represented within a diamond shape.

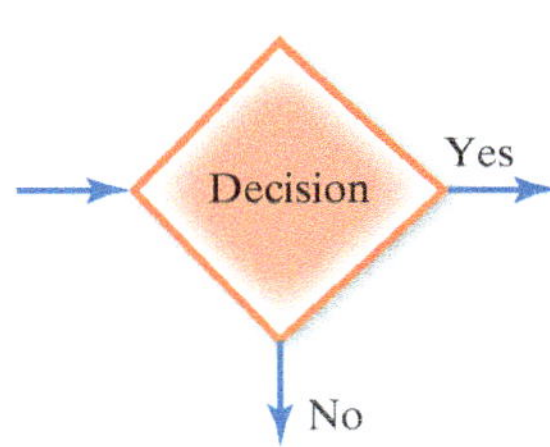

For each diamond on a flowchart, at least three arrows are always associated with the shape. The inward arrow to the diamond represents the input to the decision. It is assumed that any variables defined in the stepwise scope of a decision diamond are accessible and can be used in the decision. The two outward arrows that exit decision diamonds represent the conditional branch based on the outcome of the question asked within the diamond. If the outcome of the decision is true, the flow of the process will follow the "Yes" branch; otherwise, it will follow the "No" branch. Since no new variables are created in a decision diamond, the stepwise scope that enters the decision diamond is passed on to the next shape of each conditional branch.

EXAMPLE 15-6

Create a flowchart to express a temperature given in relative units [°F or °C] in the corresponding absolute units [K or °R].

Known:

- *Temperature in relative units (degrees Celsius or degrees Fahrenheit).*

Unknown:

- *Temperature in absolute units (kelvins or degrees Rankine).*

Assumptions:

- *Since the problem does not explicitly state the temperature to be determined, the interpreter of the algorithm will ask for the temperature and units.*

Flowchart:

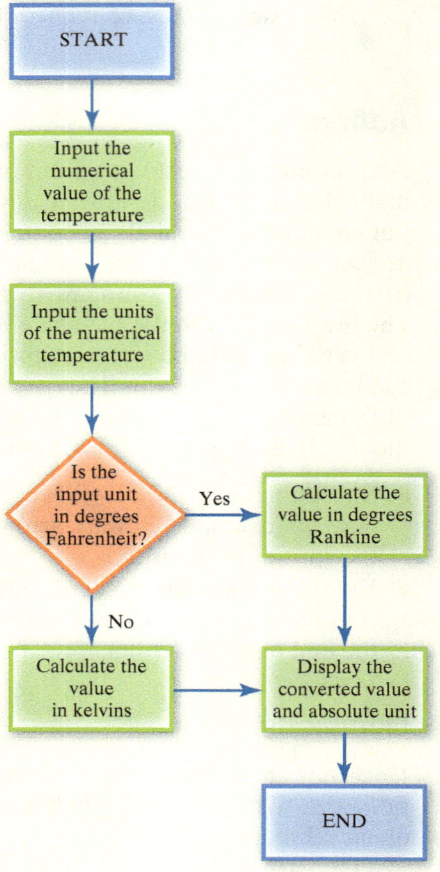

EXAMPLE 15-7

Create a flowchart to calculate the sum of a sequence of whole numbers, given the upper and lower bounds of the sequence.

Known:

- *Upper bound of whole number sequence.*
- *Lower bound of whole number sequence.*

Unknown:

- *Sum of all whole numbers between the upper and lower bound.*

Assumptions:

- *Since the problem does not explicitly state the upper and lower bounds, the interpreter will ask for the values. Include the boundary values in the summation.*

Flowchart:

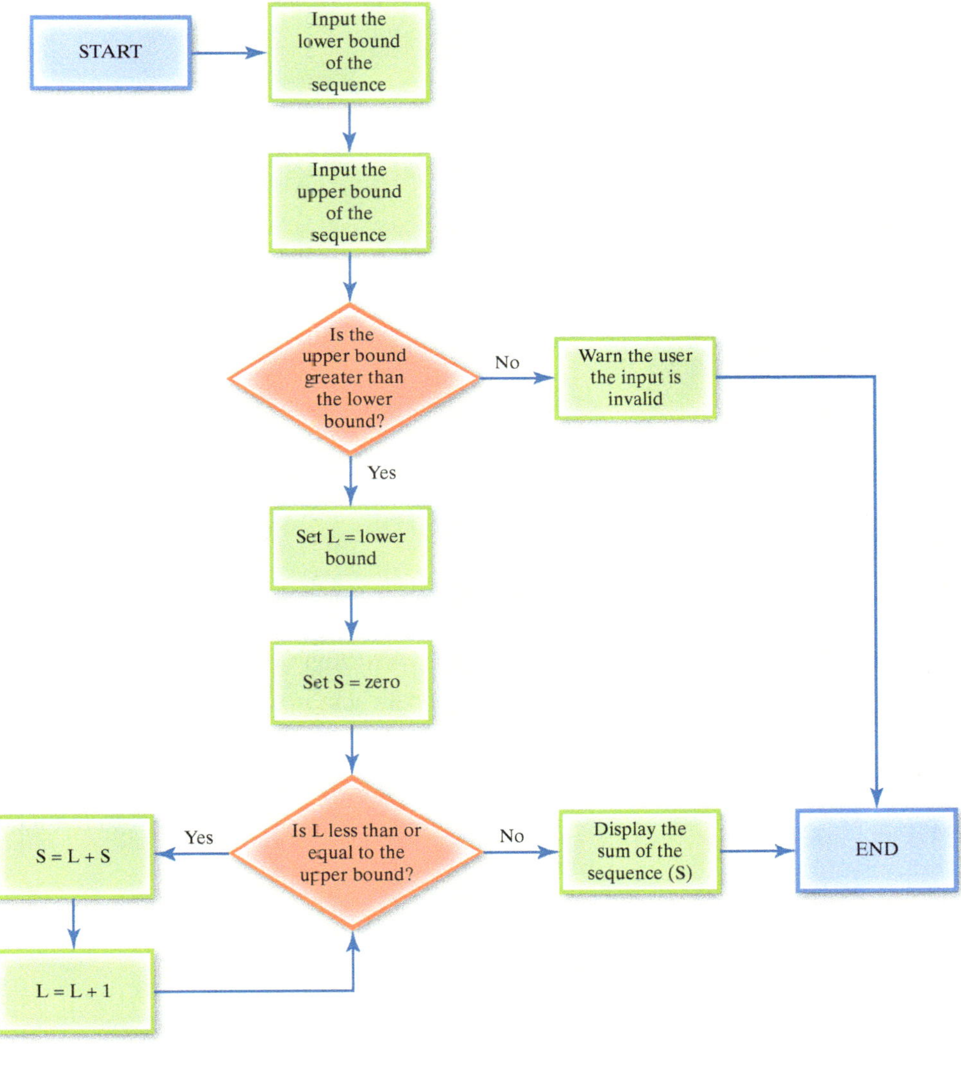

COMPREHENSION CHECK 15-4

Create a graphical algorithm to multiply all integer powers of 5, 5^x, for x between (and including) 5 and 50.

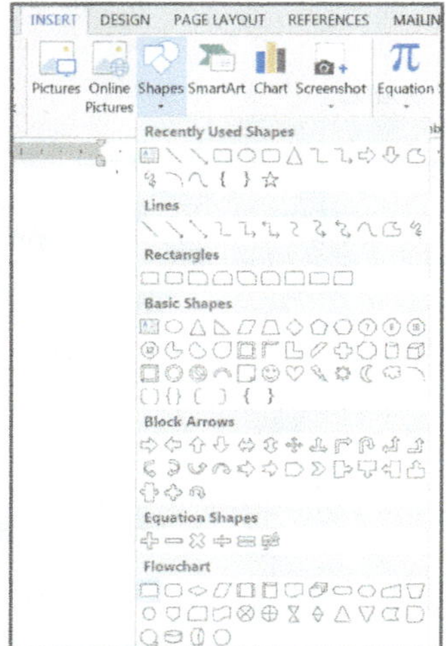

Flowchart Creation in Microsoft Word

Open a new Microsoft Word document. Click the **Insert** ribbon at the top of the Microsoft Word editor window. Click the **Illustrations > Shapes** drop-down menu. You will need to use the rectangle and diamond shapes under the Flowchart section as well as the directional arrows under the Lines section.

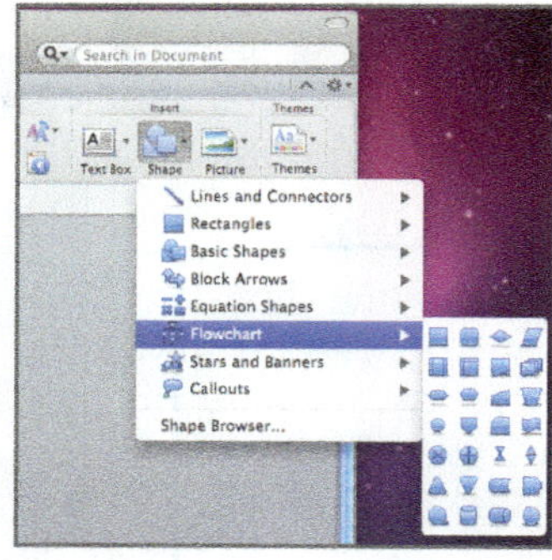

⌘ **Mac OS:** Access the flowchart shapes by clicking the Object Palette button near the top of the Formatting Palette and then clicking the Shapes button.

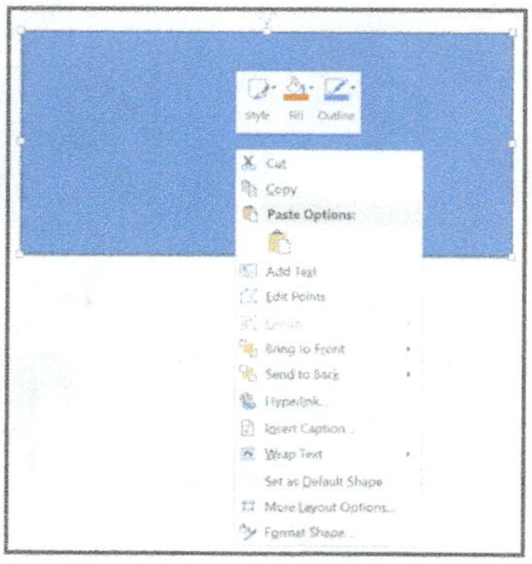

↳ **To add an action:** Click the Rectangle tool from the Shapes menu and click-and-drag into the body of the document to draw a rectangle. Right-click the rectangle and click Add Text to add text to the rectangle.

⌘ **Mac OS:** Control-click or two-finger tap to access the **Add Text** menu item.

↳ **To add a decision:** Click the Diamond tool from the Shapes menu and click-and-drag into the body of the document to draw a diamond. Right-click the diamond and click Add Text to add text to the rectangle.

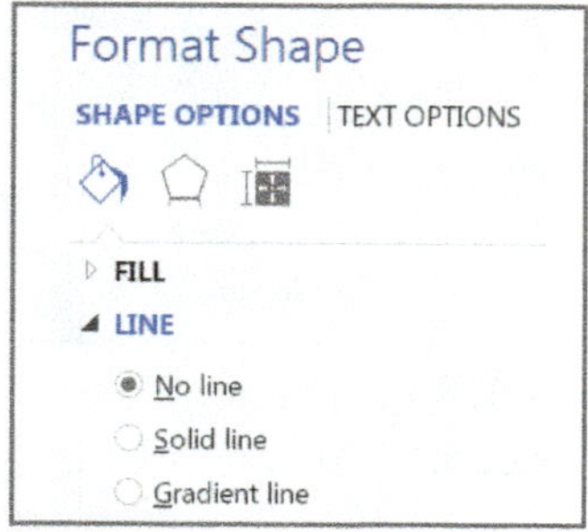

↳ **To insert the YES and NO labels:** Use the Text Box option under the Basic Shapes menu. To remove the border, right-click the text box and click Format Shape. On the Format Shape sidebar, choose **Shape Options > Paint Bucket > Line**, click "No line". Click the "X" to close the sidebar.

⌘ **Mac OS:** In the main menu, click **Insert > Text Box**. The default is probably "no border," but the border can be modified in the formatting palette.

↳ **To add an arrow:** Click the single direction arrow from the Shapes menu and starting from the source click-and-drag to the destination.

15.4 ALGORITHM BEST PRACTICES

LEARN TO: **Design algorithms that prevent unwanted results**
Utilize iteration to repeat a process a set number of times
Generate appropriate test cases for algorithms

If you have never composed a written or graphical algorithm before, the remaining part of this section details specifics on how to begin planning and writing algorithms from scratch. This section does not intend to be a definitive resource on algorithm development, but it may provide guidance if you are struggling to break down a process into small, achievable steps.

Actions

In every action within an algorithm, there must be a key verb that defines the purpose of that step within an algorithm. The remaining subsections discuss different types of actions and list some of the common verbs associated with that category of action.

Establishing Variables and Constants

After defining the scope of a problem, it might become obvious that there are intermediate calculations or assumed constants that must be contained throughout the process. Along with the explicitly defined known values, these intermediate and constant values are referred to as variables. Algorithmic variables are different from the mathematic definition of a variable because algorithmic variables are treated more like containers to store known values and results of calculations rather than being some unknown entity in a mathematical expression. They are called variables because the stored value can be written, overwritten, and used by other actions or decisions in the algorithm.

Example	**Action**
We assume the acceleration due to gravity is 9.8 meters per second squared.	**Set** variable g to be 9.8.

Other Verbs					
Set	Define	Assign	Write	Store	Designate
Label	Name	Cast	Insert	Save	Initialize

User Interaction

It is often necessary to write algorithms that can be executed with prompts for input from the person using the algorithm, provide feedback on results, or display any error messages generated in the algorithm.

User Input:

Example	Action
We want the user of the algorithm to provide the amount of water in gallons.	**Input** the amount of water in gallons, save in variable *W*.

Other Verbs					
Input	Ask	Load	Request	Query	Prompt

User Output:

Example	Action
We want the algorithm to inform the user that the amount of water can't be negative.	**Display** error message to user "Warning: amount of water can't be negative!"

Other Verbs				
Output	Display	Reveal	Write	Warn

Calculations and Conversions

When algorithms involve calculating a value using an equation, it is helpful to write out the full equation and identify which variables in the algorithm correspond to the variables in the expression. For unit conversions, it is not necessary to write out the conversion factors since those are published standards that are readily available to anyone executing your algorithm. When using conversions, it is best to list them individually so they are easily recognizable to the user. For example, when converting from feet to centimeters, the expression $L = L/3.28 * 100$ is easily recognized as the conversion from feet to meters, and then from meters to centimeters. It is harder to recognize the conversion of $L = L * 30.48$. Furthermore, it is easy to make a calculation error; it is easier to allow the program to calculate for you. When dealing with unit conversions, it is ideal to save the converted value back into the original variable to reduce the number of variables you need to keep track of in your algorithm. We will discuss MATLAB's capabilities to handle this type of equation in later chapters.

Calculations:

Example	Action
We want to calculate the thermal energy of a substance using the expression $Q = m\ C_P\ \Delta T$, where m is the mass, C_P is the specific heat, and ΔT is the change in temperature.	**Compute** the thermal energy: $Q = m\ C_P\ \Delta T$ All variables should appear in the variable list.

Other Verbs					
Calculate	Adjust	Count	Measure	Add	Multiply
Subtract	Divide	Compute	Increment	Decrement	

Conversions:

Example	Action
We want to convert a variable *t* from minutes to seconds and save the result back in the variable *t*.	**Convert** *t* from minutes to seconds, save in *t*.

Other Verbs				
Convert	Change	Alter	Revise	Switch

Referencing Other Algorithms

When developing a large program, it is sometimes helpful to break that program into several smaller programs, and then reference the smaller programs within the large program. In MATLAB, these are called **functions**. As a rule of thumb, each custom function you create should have its own separate algorithm. If you have separate algorithms for a program and the different functions referenced in the code, it makes the algorithms simpler to understand and easier to debug. When calling a function within an algorithm, it is critical to list the variables passed to the function and variables captured by the function. In general, the most common verb used with functions is "call." If you know the name you plan to use for your function, list it; otherwise, this can be set later.

Example	Action
We want to use a function named *Poltocar* that converts coordinates from polar to Cartesian. We will pass in the variable *Z* as the radius and the variable *T* as the angle. We will capture the *x*-coordinate in the variable *X* and the *y*-coordinate in the variable *Y*.	**Call** *Poltocar* In: *Z*, *T* Out: *X*, *Y*

Decisions

All decisions made in algorithms must be constructed as binary decisions. Typical decisions in algorithms involve comparing variables, examining the contents of a variable, or examining the dimensionality of a variable. Any decisions that require some amount of computation in the decision should be split so that the calculations occur in actions before reaching the decision block.

Error Checking

Including error checking in an algorithm allows for the creation of robust solutions to problems that will not lead to incorrect or unstable answers. In general, a check for an error will either terminate the algorithm or lead to some action that will allow the algorithm to continue; the program should notify the user that an error has occurred (see Figure 15-1). If you want your algorithm to re-prompt the user for input to assure that proper data are contained in a variable before proceeding into the remainder of the algorithm, see section on "Error Prevention." The remainder of this section discusses three different types of errors that may occur in an algorithm, but this is only a starting point. The amount and type of errors that can occur in an algorithm are infinite, so it is up to the designer of the algorithm to decide how and when error checking should occur.

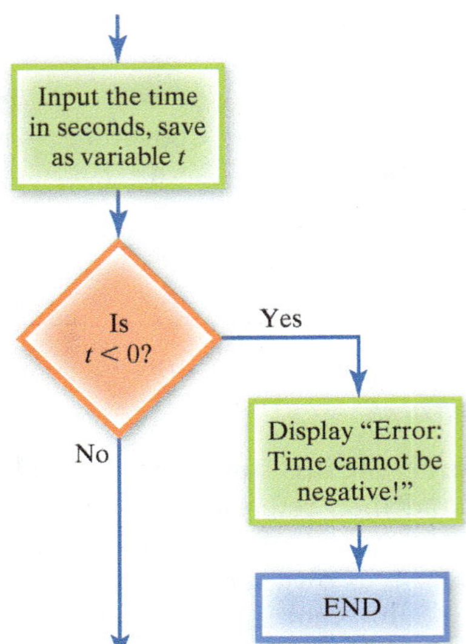

Figure 15-1 An example of error checking embedded in a flowchart.

Division by Zero and Infinite Values

If your algorithm contains a calculation where a combination of one or more variables in the computed expression could lead to a division by zero, it is smart to include a check to see if the result is zero. Some languages like MATLAB will happily compute an expression with a zero divisor and return the result as "INF"—a special MATLAB reserved word representing infinity.

NOTE

For more information on Matrix Operations, refer to Appendix A.6 online.

Invalid Dimensions of Variables

When an algorithm assumes that one or more variables contain matrices or vectors, any calculations on those variables must follow the same mathematical rules associated with the matrix operation. For example, if an algorithm requires two matrices to be added together, it would be wise to include a check to see if the two matrices have the same number of rows and columns before attempting to add them together. This will prevent algorithms from crashing due to an invalid computation. In addition, this will prevent issues related to accessing elements of a matrix that do not exist.

Invalid Range of Values

Since variables typically represent some measured or computed value, restrictions on those variables that apply in real life may not be directly enforced in your algorithm. For example, if your algorithm prompts a user to input a quantity that cannot be negative (length, volume, time, etc.), it is smart to check if the value in the variable is reasonable. Likewise, if your algorithm should not generate complex values (e.g., $3 + 2i$), your algorithm will need to check to see if the result of a computation would generate a complex value instead of the desired real value.

Error Prevention

Error prevention looks for the same type of errors that are detected in error checking, but error prevention will allow your algorithm to prompt or re-prompt the user to

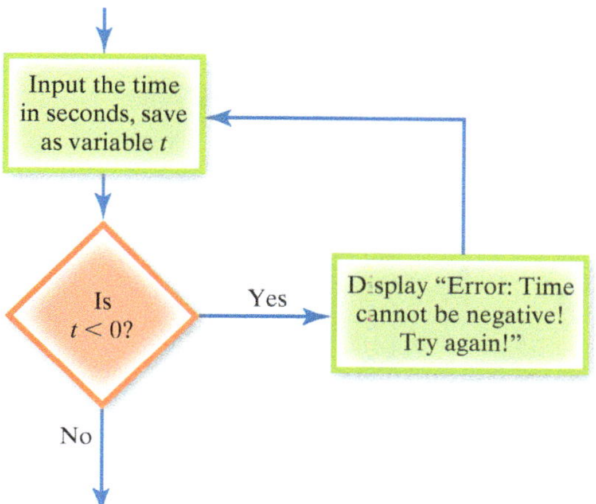

Figure 15-2 An example of error prevention embedded in a flowchart.

correct the erroneous variables. For example, if your algorithm asks the user to type in a time and the user erroneously types a negative value, your algorithm could detect the incorrect value and go back to the input statement to force the user to type the value again (and again, and again . . .) until the user types a value within the acceptable range (see Figure 15-2). In most programming languages, the structure that enables error prevention is the `while` loop.

Iteration

Some algorithms require repeated calculations that typically involve the use of a sequence of values or some operation on a vector or matrix stored in a variable. Such algorithms are considered to be iterative because they require a counter variable to keep track of when to terminate. For example, assume we have a vector, **V**, which contains positive and negative values in random order. If we want to create two new vectors, **VN** and **VP**, that contain the negative and positive values of **V** respectively, we will need to iterate through each element of the vector **V**, make a decision about each value, and store it in the corresponding vector. To do this, we would need to create a **counter variable**, or sometimes called an **index variable**, that will keep track of the number of times we have repeated a calculation or decision. If we create a counter variable, X, and initialize it to be the number 1, X will actually serve two purposes. In addition to keeping track of the number of times we have repeatedly made decisions and stored new values into **VN** and **VP**, it will also serve as the index variable into the **V** vector so that we can access element V(1), V(2), and so on until we reach the last element in **V**. Figure 15-3 demonstrates this scenario as a flowchart, including the iterated counter variable X.

Algorithms that will require iteration typically have a scenario where you have to repeat some decision or calculation "for each" or "for every" element or value within a sequence or vector. Since there is no "for each" or "for every" building block within an algorithm, this type of structure must be constructed out of the following steps:

- Initialization of a counter variable (e.g., Set X to be 1)
- A decision that involves the value of a counter variable (e.g., if X is less than or equal to the number of elements in **V**)
- Some action block that increments the counter variable (e.g., Set X equal to the current value of X plus 1)

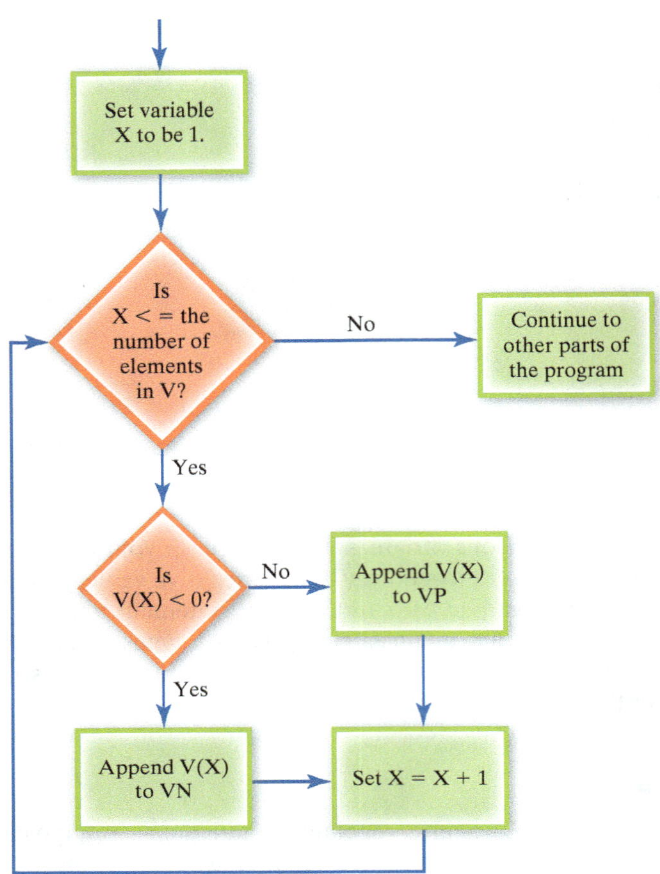

Figure 15-3 An example of iteration embedded in a flowchart.

After the counter increment action block, the algorithm can loop or refer back to the decision made on the increment variable. In most programming languages, the structure that enables iteration in an algorithm is the `for` loop.

Testing Your Algorithm

The last step in writing an algorithm is developing test cases that will reveal whether or not your algorithm behaves as expected. The key to writing test cases is figuring out how many test cases are necessary to confirm whether or not your code works. In general, there should be at least one or two test cases that will demonstrate the proper behavior of the algorithm given good input values. Not only should a test case include a list of all of the inputs used to generate the output, but you should also compute the expected output of the algorithm by hand in order to verify that the algorithm works. In addition, there should be test cases that verify all of the error checking/prevention built in to the algorithm works properly. Any decisions that lead to different states within your algorithm should have a test case to verify that the logic you designed is arranged properly in order to generate the desired output.

For example, assume we have designed an algorithm that will calculate power given energy and time. In our algorithm, we included two error checks—the first check to see if energy is greater than 50 joules and the second check to see if time is greater than 0 seconds. If either of these conditions are not true, the algorithm will display an error message "Error: Incorrect input value" and terminate. In addition, we added logic to check to see if the energy in the system is greater than 500 joules and if that is true, added 5 seconds to the time variable; otherwise, if the energy is less than or equal to 500 joules we left the time variable alone.

A test case for this scenario would look like this:

Input	Output
$E = 0$ J, $T = 30$ seconds	Error: Incorrect input value
$E = 55$ J, $T = -3$ seconds	Error: Incorrect input value
$E = 400$ J, $T = 5$ seconds	$P = 80$ W
$E = 550$ J, $T = 10$ seconds	$P = 36.7$ W

To help you develop good algorithms, we have included an algorithm template online to help you document all of the variables, procedures, and test cases necessary to create correct and verifiable algorithms. Examples of how this template can be used are provided below and on select problems in the remaining chapters.

EXAMPLE 15-8 Create an algorithm to determine the volume of a cylinder, given the radius and height.

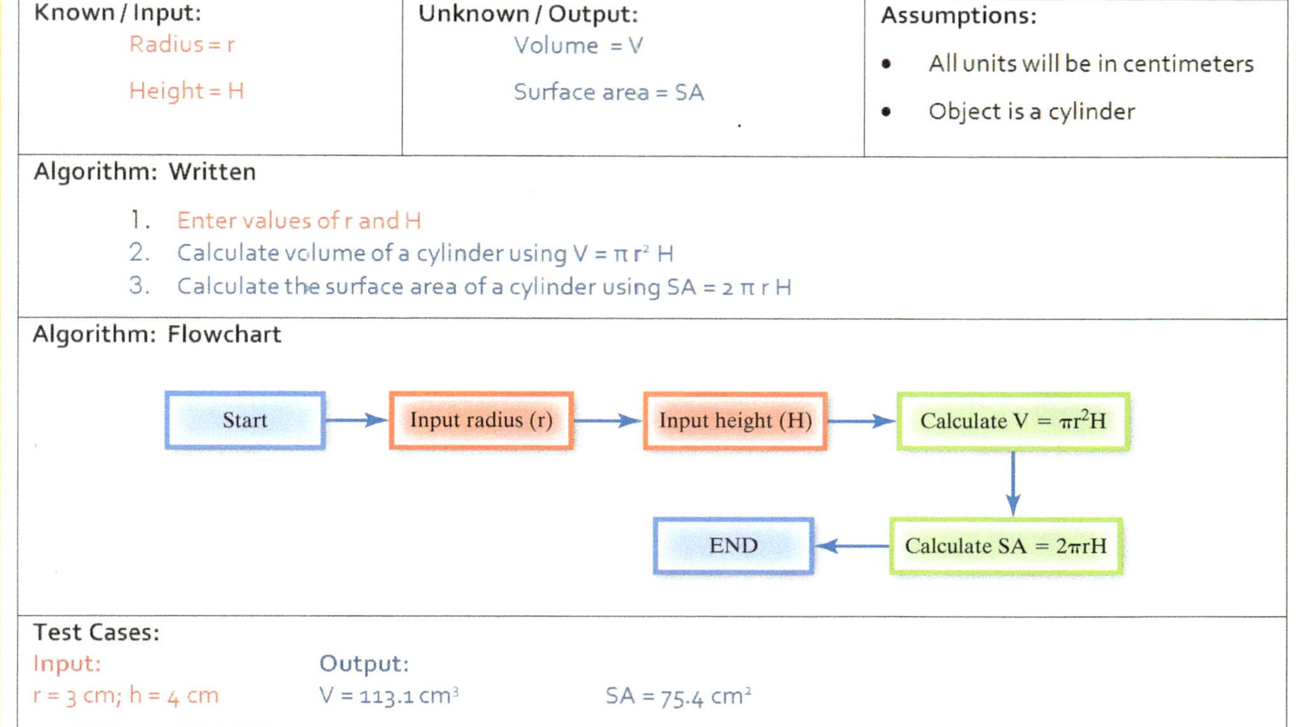

Known / Input:	Unknown / Output:	Assumptions:
Radius = r Height = H	Volume = V Surface area = SA	• All units will be in centimeters • Object is a cylinder

Algorithm: Written

1. Enter values of r and H
2. Calculate volume of a cylinder using $V = \pi r^2 H$
3. Calculate the surface area of a cylinder using $SA = 2 \pi r H$

Algorithm: Flowchart

Test Cases:

Input:	Output:	
r = 3 cm; h = 4 cm	$V = 113.1$ cm^3	$SA = 75.4$ cm^2

In-Class Activities

For ICA 15-1 to 15-7, create an algorithm (written and/or flowchart as specified by your instructor) to solve the following problems.

ICA 15-1

Your instructor will provide you with a picture of a structure created using K'Nex™ pieces. Describe the steps necessary to create the structure in the picture. When you are finished, hand your algorithm to the instructor and wait for further instruction.

ICA 15-2

Describe the steps necessary to create a paper airplane. You may assume that you are starting with a single sheet of 8½ × 11 inch paper. When you are finished, hand your algorithm to the instructor and wait for further instruction.

ICA 15-3

Describe the steps necessary to create a jelly sandwich. You may assume that you are starting with a loaf of bread, jar of jelly, a knife, and a plate on the table in front of you. When you are finished, hand your algorithm to the instructor and wait for further instruction.

ICA 15-4

An unmanned X-43A scramjet test vehicle has achieved a maximum speed of Mach number X.XX in a test flight over the Pacific Ocean, where X.XX is a positive value entered by the user. Mach number is defined as the speed of an object divided by the speed of sound. Assume the speed of sound is 343 meters per second. Determine the speed in units of miles per hour. For a test case, you may assume that the user provides the value of 9.68 for the Mach number.

ICA 15-5

Calculate a temperature provided by the user in units of Fahrenheit in units of kelvin. As a test case, you may assume the user provides the temperature of −129 degrees Fahrenheit, which is the world's lowest recorded temperature.

ICA 15-6

Determine the mass of oxygen gas (formula: O_2, molecular weight = 32 grams per mole) in a container, in units of grams. You may assume that the user will provide the volume of the container in units of gallons, the temperature in the container in degrees Celsius, and the pressure in the container in units of atmospheres. For your test case, you may assume that the user provides 1.25 gallons for the volume of the container, 125 degrees Celsius for the temperature, and 2.5 atmospheres for the pressure in the container.

ICA 15-7

Determine the length of one side of cube of solid gold, in units of inches. You may assume that the specific gravity of gold is 19.3 and that the user will provide the mass of the cube in units of kilograms. As a test case, you can assume that the user has a 0.4 kilogram cube.

ICA 15-8

The Occupational Safety & Health Administration (OSHA) defines safety regulations on working environments to protect workers from unsafe conditions. The flowchart below shows how OSHA categorizes the safety level of the working temperature given the environment temperature in degrees Fahrenheit. Given the flowchart, for what range of heat index (in degrees Fahrenheit) will the risk level be Lower (Caution), Moderate, High, or Very High to Extreme?

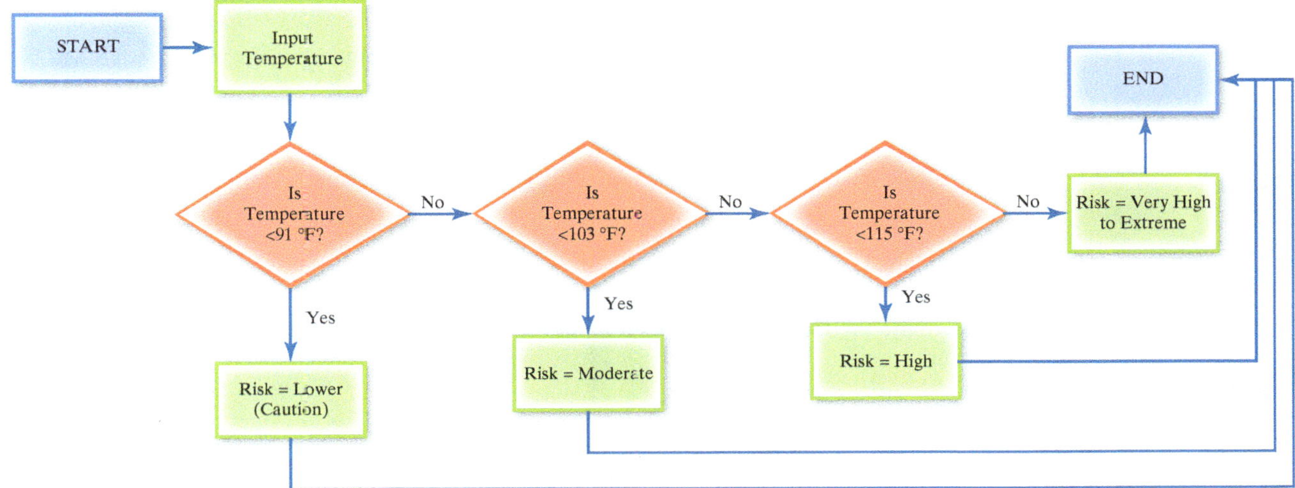

ICA 15-9

Given the flowchart below, for what range of values of pressure (in atmospheres) will the light be each of the colors Yellow, Blue, and Violet?

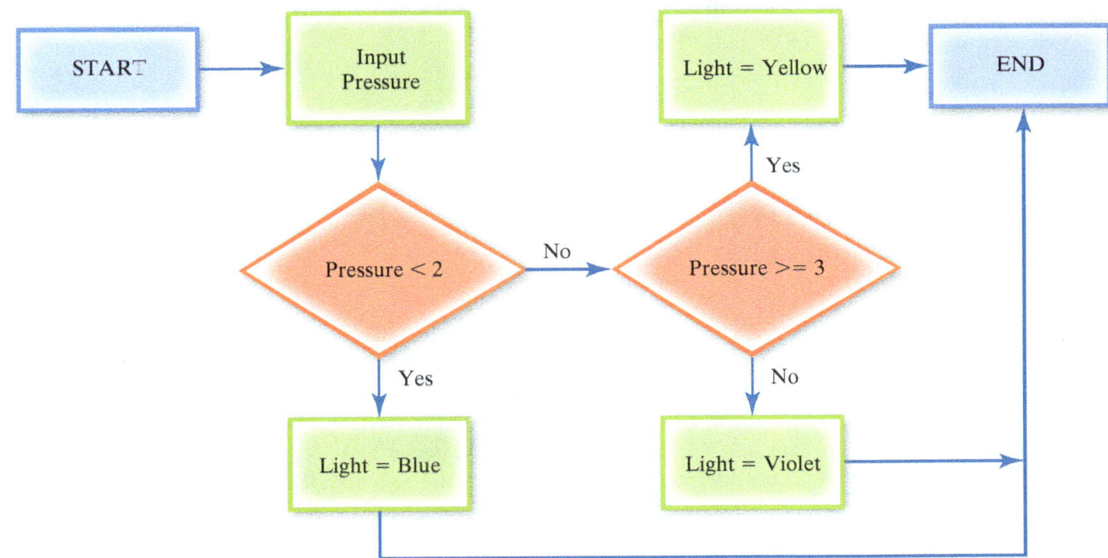

For ICA 15-10 to 15-14, create an algorithm (written and/or flowchart as specified by your instructor) to solve the following problems.

ICA 15-10

Model the decision of a driver after looking at the color of a traffic light. You may assume that the traffic light contains three bulbs (red, yellow, green) to represent stop, slow down, and go.

ICA 15-11

Describe the steps necessary to cook your favorite meal, including a starting materials list.

ICA 15-12

Describe the steps necessary to walk from your home or dorm to your engineering classroom without using any street names.

ICA 15-13

Determine the factorial value of an input integer between 1 and 10. Add a statement to check for input range.

ICA 15-14

Ask the user to draw a card from a deck of cards until the user pulls out the queen of hearts.

ICA 15-15

Create a flowchart that represents the following written algorithm.

- Input the height of person 1 [in units of feet] as P1
- Input the height of person 2 [in units of feet] as P2
- If person 1 is taller than person 2
  - Display "Person 1 is taller"
- Otherwise, if person 2 is taller than person 1
  - Display "Person 2 is taller"
  - Otherwise, display "They are the same height"

ICA 15-16

A reaction vessel is equipped with both temperature and pressure sensors. Create a flowchart that represents the following written algorithm.

- Read the values of temperature and pressure.
- If the temperature exceeds the maximum safe temperature (`TMax`), the status should be listed as "unsafe," and a variable `T` should be set to 1. Otherwise, `T` should equal 0.
- If the pressure exceeds the maximum safe pressure (`PMax`), the status should be listed as "unsafe," and a variable `P` should be set to 1. Otherwise, `P` should equal 0.
- If both temperature and pressure are less than their corresponding maxima, the status should be listed as "safe" and both variables `T` and `P` should be set to 0.
- Repeat the entire procedure.

Chapter 15 REVIEW QUESTIONS

1. Create an algorithm to determine the weight of a rod in units of pound-force on the surface of Callisto, Jupiter's moon. You may assume that the user of the algorithm will provide the volume of the rode in units of cubic meters. Your algorithm may assume that the specific gravity is 4.7 and the gravitational acceleration on Callisto is 1.25 meters per second squared. As a test case, you can assume the user has a rod with a volume of 0.3 cubic meters.

2. The Eco-Marathon is an annual competition sponsored by Shell Oil, in which participants build special vehicles to achieve the highest possible fuel efficiency. The Eco-Marathon is held around the world with events in the United Kingdom, Finland, France, Holland, Japan, and the United States.

   A world record was set in the Eco-Marathon by a French team in 2003 called Microjoule with a performance of 10,705 miles per gallon. The Microjoule runs on ethanol. Create an algorithm to determine how far the Microjoule will travel in kilometers given a user-specified amount of ethanol, provided in units of grams. For your test case, you may assume that the user provides 100 grams of ethanol.

3. Create an algorithm to determine the density of tribromoethylene in units of kilograms per cubic meter in a cylindrical tank. Your algorithm should be written in such a way that the user provides the height measured in units of feet of the tribromoethylene in the tank, the surface pressure in units of atmospheres, and the total pressure at the bottom of the tank, also measured in atmospheres. For your test case, you may assume that the user provides a height of 25 feet for the tank, tank surface pressurization of 3 atmospheres, and a total pressure at the bottom of the tank of 5 atmospheres.

4. Create an algorithm to determine how long, in units of seconds, it will take a motor to raise a load into the air. Assume the user will specify the power of the motor in units of watts, the rated efficiency as a percentage (in whole number form—for example, 50 for 50%), the mass of the load in kilograms, and the height the load is raised in the air in units of meters. As a test case, you may assume the user provides the 100 watts for the power of the motor, 60% for the efficiency of the motor, 100 kilograms for the mass of the load, and 5 meters for the height the load is raised.

5. Create an algorithm to determine how long it will take to boil acetone using a hot plate, assuming acetone has a boiling point of 56 degrees Celsius. You may assume that the user of your algorithm will need to ask for the amount of time (split into the number of minutes and seconds) it takes to boil a user provided volume of water, measured in milliliters, as well as the temperature of the lab in degrees Fahrenheit, which is also the initial temperature of the water before boiling. In addition, the user will provide the power rating of the hot plate in units of watts and the volume of acetone to be boiled in the experiment in milliliters. Note that your algorithm must determine the efficiency of the hot plate. As a test case, the hot plate takes an average of 8 minutes and 55 seconds to boil 100 milliliters of water in a 75 degree Fahrenheit laboratory with a 283 watt hot plate and it is desired to boil 100 milliliters of acetone.

6. The Apple TV™ is a personal video device created by Apple, Inc that attaches to high-definition televisions to present streaming content to a viewer on demand, in contrast to a standard source (cable/satellite) set-top box or digital video recorder (DVR) that presents recorded or live broadcast signals on a television. According to a study conducted by the Natural Resources Defense Council in June 2011, if consumers were to move from set-top boxes to smaller "thin" devices like the Apple TV, there would be an overall 70% reduction in annual energy costs resulting from use of set-top boxes if they were replaced by smaller energy efficient devices like the Apple TV.

   Below is a data table that shows the typical power consumption of the Apple TV in different power states (standby/off, idle—not playing any video content, playing via Ethernet, and playing via Wifi). Create an algorithm that allows a user to provide the power currently being used by their Apple TV in units of watts and determine the state of the device. If the user provides value for the power outside of the range in the table below, your algorithm should report that the state of the device is unknown.

Power Consumption of the Apple TV

State	Power Consumption [W]
Off/Standby	less than 0.5
Idle	0.5 to 1.5
Streaming via Ethernet	1.5 to 1.6
Streaming via Wifi	1.6 to 2

7. Humans can see electromagnetic radiation when the wavelength is within the spectrum of visible light. Create an algorithm to determine if a given wavelength [nanometer, nm] is one of the six spectral colors listed in the chart below. Your algorithm should ask the user to enter a wavelength, then indicate in which spectral color the given wavelength falls or provide a warning if it is not within the visible spectrum.

Color	Wavelength Interval [nm]
Red	~ 700–635
Orange	~ 635–590
Yellow	~ 590–560
Green	~ 560–490
Blue	~ 490–450
Violet	~ 450–400

8. Create an algorithm to display the letter grade in a course. A course grade is typically reported as a real number with values from 0 to 100. Your algorithm should ask the user to provide a grade in numerical form and display the letter grade corresponding to a numerical grade if the entered value is between 0 and 100 inclusive. If the value is outside this range, ask the user to enter the value again.

Typical range of grades:

A: 90 <= grade
B: 80 <= grade < 90
C: 70 <= grade < 80
D: 60 <= grade < 70
F: grade < 60

9. Create an algorithm to determine whether a given altitude [meters] is in the troposphere, lower stratosphere, or the upper stratosphere. The algorithm should include a check to ensure the user entered a positive value. If a non-positive value is entered, the program should warn the user and terminate. In addition, your algorithm should calculate and report the resulting temperature in units of degrees Celsius [°C] and pressure in units of kilopascals [kPa]. Refer to atmosphere model provided by NASA: http://www.grc.nasa.gov/WWW/K-12/airplane/atmosmet.html.

10. Create an algorithm to determine whether a given Mach number is subsonic, transonic, supersonic, or hyper sonic. Assume the user will enter the speed of the object in meters per second, and the program will calculate the Mach number and determine the appropriate Mach category. As output, the program will display the Mach catagory. The algorithm should include a check to ensure the user entered a positive value. If a non-positive value is entered, the program should warn the user and terminate. Refer to the NASA page on Mach number: http://www.grc.nasa.gov/WWW/K-12/airplane/mach.html.

11. Construct an algorithm to determine the class rank (freshman, sophomore, junior, or senior) of students enrolled in this course. You may assume the user will enter the total number of students in the course, and the program will ask the user to enter the total number credit hours earned by each student. The program will count the number of students in each class rank. The program will produce as output the total number of students in the class and the percentage of students who are in each class rank.

In this algorithm, you may assume that a freshman has earned less than 30 credit hours, a sophomore between 30 and 59, a junior is between 60 and 89, and a senior has earned 90 or more credit hours.

12. Create an algorithm to ask the user to enter the number of Internet-enabled devices they own. If the user enters a negative number or a number greater than 50, the message "This is not reasonable. Try again." should appear to the user and they should be asked to enter another value. The error message should continue to appear each time they enter an invalid value. When the user provides an acceptable value, the message "Thank you" should be displayed and the algorithm should terminate.

13. Create an algorithm to first ask the user to enter a positive number. After checking that the value entered is positive, determine all of the even numbers between 0 and that value that are also evenly divisible by the number three. If the user enters a non-positive value, ask the user to enter a new value. At the end of your algorithm, your algorithm should display the total number of values that meet the specified criteria.

HINT: The mathematical operator modulus (%) will return the remainder after division by a specific number. For example, 5% 4 is 1 because 5 can be evenly divisible by 4 once, with a remainder of one—the modulus. Likewise, 10% 4 is 2 because 10 can be evenly divisible by 4 twice, with a remainder of two—the modulus.

14. Create an algorithm to ask the user to enter a positive value. You may assume the user enters a positive number and you do not need to check if the value is indeed positive. Determine which is greater: the sum of the whole-even numbers between 0 and the entered value or the sum of the whole-odd numbers between 0 and the entered value. At the end of your algorithm, your algorithm should display "The sum of the odd numbers is greater" or "The sum of the even numbers is greater", depending on the outcome of the algorithm.

15. You want to calculate your grade point ratio (GPR). Create an algorithm to input your courses for this semester, the semester hours for each course, the letter grade you expect to earn, and number of grade points earned for each course (such as A = 4). At the end of the algorithm, your algorithm should display the average GPR for the semester using the following formula: summation of the product of the semester hours and number of points earned divided by the summation of semester hours.

CHAPTER 16

MATLAB VARIABLES AND DATA TYPES

A variable in MATLAB is not like a variable in mathematics. In math, a variable is typically an unknown for which you wish to determine a value. In MATLAB, on the other hand, a variable is simply a label for a container (or location) in which information is stored. Variables can be classified according to the type of information they hold. In the workspace, the data type of each variable is represented by an icon shown in the leftmost column of the workspace by the variable name. Table 16-1 below illustrates the different variable classifications in MATLAB.

Table 16-1 Variable Classifications in MATLAB

ICON	Data Type	Description
	Numeric Array	Scalars, vectors, and matrices containing numbers
abc	Character Array	Text strings. Each character occupies one element
✓	Logical Array	0 means false; 1 means true Discussed in Chapter 19
{ }	Cell Array	Contains mixed data types. Indexed numerically.
	Structure Array	Contains mixed data types. Data organized into named fields.
	Sparse Array	Saves memory by storing only non-zero elements
	Symbolic Array	Allows symbolic mathematics within MATLAB Not covered in this text

16.1 VARIABLE BASICS

LEARN TO: Remember the naming rules for variables
Understand the use of = as the assignment operator

Every variable in MATLAB must have a name. This allows us to keep track of large amounts of information and gives a means whereby our code can use the stored information relatively easily. MATLAB names must follow certain rules to avoid unintended side-effects.

Naming Rules

- All names must consist only of letters, numbers, and the underscore character (displayed by holding "Shift" and typing a hyphen).
- Names can *only* begin with alphabetic letters (no numbers or special characters).
- Names cannot be longer than 63 characters.
- Names in MATLAB are case sensitive: `Frog` and `frog` are different.
- Names cannot have the same name as any other identifier (program name, function name, built-in function, built-in constant, reserved word, etc.), because MATLAB's order of execution will prevent the program from finding the correct item.

These rules apply to ANY named item in MATLAB, not just variables. We will give further detail in the next chapter.

Always choose names that have meaning within the context of the program, not just random letters. We cannot overemphasize the importance of well-chosen variable names in creating self-documenting code: code that explains itself. For example, if a variable contained the diameter of a bolt, the name `BoltDiam` is immediately recognizable for what it contains; a randomly chosen variable such as `X1` or `Q` has no meaning in the context of the program.

COMPREHENSION CHECK 16-1

Which of the following are valid MATLAB variable names? For those that are not valid, explain why they are invalid.

(a) `m`	(c) `4mass`	(e) `MyFile-docx`
(b) `mass6`	(d) `m&m`©	(f) `ReactorYield`

The Assignment Operator

Before beginning our discussion of data types in MATLAB, we need to introduce the concept of the **assignment operator**. The purpose of an assignment operator is to tell the computer to put a value into a variable. Perhaps unfortunately, the most commonly used assignment operator in modern programming languages is =, or the "equal" sign. We say unfortunately, because this notation often confuses novices. In mathematics, a statement such as `A=A+1` is an equation and has exactly two solutions for A: infinity and negative infinity. However, in programming this means "add one to whatever is in variable A, then place the result back in A.

To avoid the dual problem of mistaking = for "equals" and to explicitly indicate the direction of the movement of information, some languages use a left pointing arrow (or other symbols) for the assignment operator.

It is important to note in MATLAB the = symbol should be read as "is assigned the value of". Thus, the programming statement `V2=V2+4` is read "V2 is assigned the value of (the original value of) V2 plus four". Note that the original value of V2 is lost in this process.

It is also VERY important to note that the transfer of data in an assignment statement is ALWAYS from right to left.

Consider the programming statement `Ex1=V3-V4^P1`.

When the computer executes this statement, it first recognizes that it is an assignment statement. Therefore it first looks at the expression to the right of the equals and determines its value: Raise the value in `V4` to the value in `P1` and subtract that from the value in `V3`. Next, it looks to the left of the equal sign to find the destination, thus placing the calculated result into the variable `Ex1`.

It is also important to understand that whatever appears to the left of the assignment operator MUST be a variable, or as we will see later, a group of variables. For example, you will get an error if you type `A+ 1=A` since you cannot have a computation on the left of the equal sign. Also, any variables that appear on the right must already be defined before the statement is executed.

COMPREHENSION CHECK 16-2

Which of the following assignment statements are valid? For those that are not valid, explain why. For those that are valid, show what value is placed into which variable. Assume the following variables (and no others) have already been defined: `A=2 B=-1 C=3 D=2.5`

Note that the priority of operators is the same as in Microsoft Excel.

(a) `X=A`
(b) `B=Y`
(c) `Z=A+55`
(d) `W=A^B+C`
(e) `V=D-Q^2`
(f) `U+13=65`
(g) `T^2=4`
(h) `Part#3=C`
(i) `D=D^B`
(j) `A=9999`
(k) `A+B=C-5`
(l) `QPDoll=C^C`

16.2 NUMERIC TYPES AND SCALARS:

LEARN TO: Define different numeric types in MATLAB
Perform basic scalar operations
Write mathematic expressions using scalar functions

Numeric data is exactly that—information consisting of one or more numbers. In MATLAB, as in most computer languages, there are actually several different types of numeric data. As an example, numbers can be classified as integers and non-integers (floating point numbers), and the two types are encoded differently using binary digits (1s and 0s) inside the computer. The major numeric types are listed in Tables 16-2 and 16-3, but a complete discussion is beyond the scope of this introductory material. For the moment, it is sufficient to note that the default numeric type in MATLAB is double precision, even if the value being stored has no fractional part.

Since the default numeric type in MATLAB is double precision, and doubles have the greatest range and, except for `int64` and `uint64`, also the greatest precision of all of the standard numeric types, why would you ever want to change to another format with less precision and/or range? Occasionally, when your application is processing gargantuan quantities of data, switching to either single precision or one of the integer formats will save memory space and/or increase execution speed. Since memory is cheap and processors are amazingly fast, this is seldom a significant factor in most engineering applications, and is not worth dealing with the possible problems of increased round-off error, results beyond the range of the selected format, etc.

In addition to numeric types, variables containing numbers can also be categorized based on how many values are stored in that variable, and how they are arranged. MATLAB variables containing only numeric data are considered to be matrices, hence the name MATrix LABoratory. We will separate these into three broad categories: scalars, vectors, and matrices. We will discuss vectors and matrices further in Sections 16-3 and 16-4.

Table 16-2 Floating Point Formats

Name	Bytes	Approximate Precision	Non-zero Minimum	Non-infinite Maximum
Double	8	16 decimal digits	$\pm 2.2251 \times 10^{-308}$	$\pm 1.7977 \times 10^{308}$
Single	4	7 decimal digits	$\pm 1.1755 \times 10^{-38}$	$\pm 3.4028 \times 10^{38}$

Table 16-3 Integer Formats

Bytes	Signed			Unsigned		
	Name	Minimum	Maximum	Name	Minimum	Maximum
1	`int8`	−128	127	`uint8`	0	255
2	`int16`	−32,768	32,767	`uint16`	0	65,535
4	`int32`	−2,147,483,648	2,147,483,647	`uint32`	0	4,294,967,295
8	`int64`	-9.2234×10^{18}	9.2234×10^{18}	`uint64`	0	1.8447×10^{19}

Scalars

In this section we will discuss scalars and a variety of issues relating to scalars, although many of the concepts will carry over to our discussion of vectors and matrices.

A **scalar** is simply a single numeric value, such as $7, -35.12, 2.778 \times 10^{15}$, or $2.4 + 3.92i$. **In MATLAB, a scalar is actually considered a 1 × 1 matrix.**

To define a scalar, you simply use the assignment operator (=) to place the desired value into a variable. Remember that the location into which a value is being place is ALWAYS to the left of the equal sign.

Examples of scalars include:

```
Num=34
Age=62
Root1=15.67-14.65i
BigNum=3.5E246
TinyNum=5.7E-105
```

Predefined Constants

The following names are already defined as special constants in MATLAB. You should avoid using them for other purposes.

pi = the constant pi(π) to 15 decimal places

i = $\sqrt{-1}$

j = $\sqrt{-1}$

COMPLEX NUMBERS

```
Q=sqrt(-9);
```

Q will contain 0 + 3i where $i = \sqrt{-1}$

NOTE: j is used by electrical engineers to avoid confusion with electric current which usually uses i as the variable.

COMPREHENSION CHECK 16-3

Write MATLAB code to complete the following commands.

(a) Place the number 8 into the variable `Int2`.
(b) Place the number thirty five point seven into the variable `Real2`.
(c) Place the number 47.98×10^{56} into the variable `Big2`.
(d) Place the number 3×10^{-15} into the variable `Small2`.

Calculations with Scalars

For the most part, the mathematical operators used in Microsoft Excel work the same way in MATLAB, including parentheses and priority of operators.

EXAMPLE 16-1

What is stored in the variable when each line of MATLAB code below is exectuted?

```
DPM=38/19+5*3;
```

`DPM` *contains 17*

```
RTP=3^2+1;
```

`RTP` *contains 10*

```
RTPP=3^(2+1);
```

`RTPP` *contains 27*

Recall variable names should have meaning in the context of their use. Here, `DPM` refers to Divide Plus Multiply, `RTP` refers to Raise To Power, and `RTPP` refers to Raise to Power with multiple elements in the Power.

Numeric Functions

There are hundreds of functions built into MATLAB for performing a wide variety of operations with numeric data. These are similar to the functions you learned about in your study of Excel, and in many cases even have the same name, such as `sin`, `sqrt`, or `round`. We will introduce quite a few functions throughout the subsequent sections. A sample of functions used with numeric values is shown in Table 16-4.

Table 16-4 Selected Numeric Functions

Function	Description	Function	Description
`single`	Convert to single precision	`round`	Round to nearest integer
`double`	Convert to double precision	`ceil`	Round toward positive infinity
`int8`	Convert to 8 bit integer	`floor`	Round toward negative infinity
`int32`	Convert to 32 bit integer	`fix`	Round toward zero
`uint64`	Convert to unsigned 64 bit integer	`sqrt`	Determine the square root
`inf`	Infinity	`nthroot`	Determine the Nth root
`NaN`	Not a Number	`sin`	Sine of angle in radians
`intmax`	Maximum for specified integer type	`sind`	Sine of angle in degrees
`intmin`	Minimum for specified integer type	`acos`	Inverse cosine in radians
`realmax`	Maximum for specified floating-point type	`acosd`	Inverse cosine in degrees
`realmin`	Minimum for specified floating-point type	`exp`	Exponential function
`real`	Real part of a complex number	`log`	Natural logarithm
`imag`	Imaginary part of a complex number	`log10`	Base 10 logarithm

Functions to Help You Learn Other Functions

The two functions `help` and `doc` can be invaluable in learning what various MATLAB functions do and how they are used (their syntax). They can also help you explore the often labyrinthine sets of built-in functions to discover new and useful features of the language.

To illustrate the use of the `help` function, at the prompt in the command window type `help log`. A short explanation of log appears, along with the syntax used. Near the end of the information is a section that says "See also" followed by a list of hyperlinks to related functions.

`doc` and `help` in general should only be used as single line commands in the Command Window, not included in an m-file. M-files will be discussed in the next chapter.

To illustrate the use of the `doc` function, at the prompt in the Command Window type `doc weekday`. This launches the documentation browser. The documentation browser makes it relatively easy to explore MATLAB. In the figure below, note the following features:

- The **Contents** outline on the left. This shows an outline of the organization of MATLAB functions into related groups, and includes drop-down menus for Examples, Concepts, Troubleshooting, etc. If the Contents outline is not visible, click the little outline icon near the top left to expand it.
- Hyperlinks to examples on the left of many subtopics within the main informational window.
- The search box at the top of the main information window. Here, you can type keywords to search for functions that do particular types of tasks.

- Scrolling down to the bottom of the main information window will list related functions under **See Also**, similar to the help function.

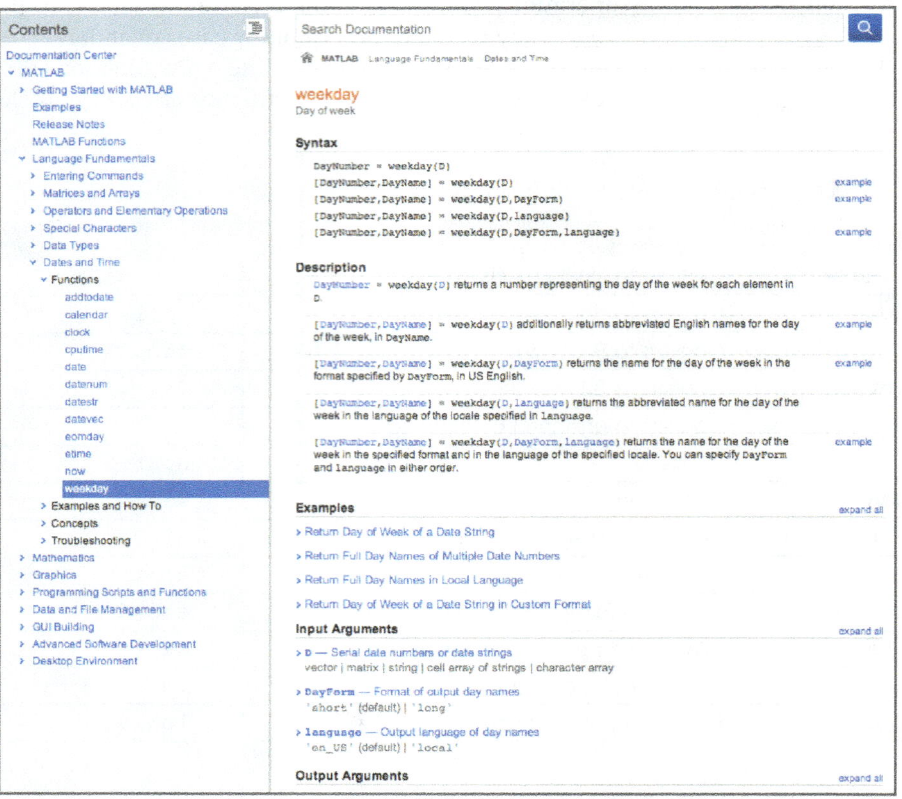

Displaying Numbers

The default numeric display format in MATLAB is to show four decimal places unless the value has no fractional part. For numbers with fractional parts and a magnitude less than 10,000 but greater than 0.001, MATLAB will display up to three digits to the left of the decimal point and four decimal places. For values outside of this range, MATLAB will display the value in scientific notation, with the most significant digit to the left of the decimal point and four decimal places.

`100*pi`	will display as 314.1593
`1000*pi`	will display as 3.1416e+03
`pi/1000`	will display as 0.0031
`pi/10000`	will display as 3.1416e–04

Values with no fractional part and a magnitude less than 10^{10} will display as an integer. Larger values will display with five significant figures in scientific notation as above.

`123123123`	will display as 123123123
`1231231231`	will display as 1.2312e+09

NOTE

Even if a number has no fractional part, it will be stored as a double precision floating point number. If you want a value to be stored in one of the integer formats, you must specifically tell MATLAB to do so.

You may change the default display format with the `format` function. For example `format long` will change the default display format to show 15 digits instead of only 5. More information about the format function can be found by typing `help format` or `doc format` in the Command Window.

COMPREHENSION CHECK 16-4

(a) Calculate the two roots of the quadratic equation $2x^2 + 2x + 1 = 0$ using the quadratic formula and place them in `R1` and `R2`. Note that MATLAB should do this calculation—do not calculate by hand and enter the resulting values.

(b) Calculate the tangent of 75 degrees and place the result in `Trig2`.

16.3 VECTORS

LEARN TO: Define numeric vectors in MATLAB's workspace
Perform basic vector operations
Write mathematic expressions using functions on an entire vector

A **vector** consists of numeric values organized into a single row or a single column. Examples include $[3.5 \quad -38 \quad 0 \quad -1]$ and $\begin{bmatrix} -1 \\ 0 \\ 3.2 \end{bmatrix}$. **A vector is either an N × 1 matrix (column vector) or 1 × N (row vector), where N is an integer greater than 1.**

Defining Vectors

As with scalars, the usual method for defining a vector is to use the assignment operator (=). There are a few immediate questions, however, since vectors have more than one value and may be organized as either a row or column.

1. How do we tell MATLAB we are defining a vector and not a scalar?
2. How do we tell MATLAB whether we are creating a row or a column vector?
3. How do we distinguish the individual values from each other?

First, when we wish to define a vector, the first character following the equal sign is [(open square bracket). Once we have entered all of the numbers, we end with] (close square bracket). It is also possible to define vectors using indexing, which is discussed later in this section.

NOTE

A **space** or **comma** delimiter creates a row

A **semicolon** delimiter creates a column

The solution to the second question also solves the third: we separate the numbers by **delimiters**—either a space, a comma, or a semicolon. If we separate the individual values with spaces or commas, a row vector will be created. If we separate the values with semicolons, they will be placed in a column vector.

EXAMPLE 16-2

What is stored in the variable when each line of MATLAB code below is exectuted?

```
Pair= [23 4.3];
```

`Pair` *contains the row vector* `[23 4.3]`

```
Trio=[1, 4, 79857];
```

`Trio` *contains the row vector* `[1 4 79857]`

```
CTrio= [0;0;97];
```

`CTrio` *contains the column vector* $\begin{bmatrix} 0 \\ 0 \\ 97 \end{bmatrix}$

The Transpose (') Operator

If you need to convert a row vector to a column vector, or vice versa, use the apostrophe or single quote to transpose it.

Consider the following two lines of code:

```
V=[6,-9,3];          VT=V';
```

V will contain $[6 \quad -9 \quad 3]$ VT will contain $\begin{bmatrix} 6 \\ -9 \\ 3 \end{bmatrix}$

Special Case: Linear Sequences

It is fairly common to need a vector containing a set of equally spaced values, so MATLAB includes a short-cut method for creating such sequences.

If you wish to create a sequence of values in a row vector incrementing by one, such as 1, 2, 3, 4, 5, you can use the **colon operator** to separate the first and last value of the sequence.

NOTE

A **colon** delimiter creates a sequence

EXAMPLE 16-3

What is stored in the variable when each line of MATLAB code below is exectuted?

```
LSeq=1:5;
```

`LSeq` *contains* `[1 2 3 4 5]`

Note that the brackets are optional in this case. You could also enter `LSeq=[1:5];`

If you wish a sequence that increments by a value other than 1, you may place the increment value between the first and last value, separated by colons.

EXAMPLE 16-4

What is stored in the variable when each line of MATLAB code below is exectuted?

```
LseqA=5:15:80;
```

`LSeqA` *contains* `[5 20 35 50 65 80].`

```
Frac=0.7:0.03:0.88;
```

`Frac` *contains* `[0.7 0.73 0.76 0.79 0.82 0.85 0.88]`. *Fractional values are allowed*

```
Rev=4:-3:-2;
```

`Rev` *contains* `[4 1 −2]`. *Negative values are allowed.*

```
Miss=1:5:20;
```

`Miss` *contains* `[1 6 11 16]`. *Note that the increment did not land on the final value (20). In cases like this, the sequence will stop on the last value BEFORE it exceeds the stated end value.*

COMPREHENSION CHECK 16-5

Each problem should be done with a single MATLAB statement:

(a) Create a column vector named `N4` containing the values 17, 34, −94, 16, and 0.
(b) Create a row vector named `Tiny` containing the values 3.4×10^{-14}, 9.02×10^{-23}, and 1.32×10^{-9}
(c) Create a row vector named `Ev` containing all even integers from 2 to 250.
(d) Create a column vector name `Tenths` containing the decreasing sequence 10, 9.9, . . . 0.2, 0.1, 0

Calculations with Vectors

Basic calculations with vectors are somewhat more complicated than the same calculations with scalars.

Addition and Subtraction

When addition or subtraction involves a vector, there are two cases:

1. A scalar and a vector: the scalar is added or subtracted to each element of the vector.
2. Two equal size vectors: corresponding elements of the two vectors are added or subtracted.

If you attempt to add two vectors that are not the same size, you will get angry red letters, indicating an error.

EXAMPLE 16-5

What is stored in the variable when each line of MATLAB code below is exectuted?

Assume `V1=[2 5 9];` `V2=[3 0];` `V3=[4 -7 0];`

```
D1=V1-4;
```

`D1` *contains* [−2 1 5].

```
S1=V1+V3;
```

`S1` *contains* [6 −2 9].

```
E1=V1+V2;
```

Will return the message: `Error using + Matrix dimensions must agree.`

Multiplication and Division

A complete discussion of vector multiplication will be deferred to the next section on matrices. For now, we will only consider cases similar to the two mentioned for addition and subtraction:

1. A vector multiplied or divided by a scalar: each element of the vector is multiplied or divided by the scalar. NOTE: if you attempt to divide a scalar by a vector, you will get angry red letters.
2. A vector multiplied or divided by an equal length vector: corresponding elements of the two vectors are multiplied or divided. NOTE: This requires the use of the **element-wise operator**, sometimes referred to as the "**dot operator**" since the actual operator is preceded by a period (or "dot") symbol, as shown below.

NOTE

The **dot operator** indicates an element-wise operation

If you attempt an element-wise multiplication or division of two vectors that are not the same size, you will get angry red letters.

EXAMPLE 16-6

What is stored in the variable when each line of MATLAB code below is exectuted?

Assume `V1=[2 5 9]; V2=[3 0]; V3=[4 -7 0];`

```
Q1=V1/5;
```

`Q1` *contains* `[0.4 1 1.8]`.

```
P1=V2*300;
```

`P1` *contains* `[900 0]`.

```
P2=V1.*V3;
```

`P2` *contains* `[8 −35 0]`. *Note the dot before the asterisk (.* instead of *). This tells MATLAB that the multiplication is to be done on an element-wise basis.*

```
Q2=V3./V1;
```

`Q2` *contains* `[2 −1.4 0]`. *Note the dot before the slash (./ instead of /). This tells MATLAB that the division is to be done on an element-wise basis.*

```
Q3=5/V1;
```

Will return the message: `Error using / Matrix dimensions must agree.`

```
P3=V1*V3;
```

Will return the message: `Error using * Matrix dimensions must agree.`

The reasons for these errors will become clear when we discuss matrices in the next section.

```
Q4=V3/V1;
```

`Q4` *contains* −0.2445. *For an explanation of this seemingly mysterious result, refer to any general reference on matrix operations. For the moment just take note that vector division without the dot probably does not do what you expected.*

Powers

We will consider three cases here, all of which are element-wise operations.

1. A vector raised to a scalar.
2. A scalar raised to a vector.
3. A vector raised to a vector of equal size.

In all three cases, we will use .^ (not simply ^) to inform MATLAB to do an element-wise operation.

EXAMPLE 16-7

What is stored in the variable when each line of MATLAB code below is exectuted?

Assume `V1=[2 5 9]; V2=[3 0]; V3=[4 -7 0];`

```
R1=V1.^3;
```

`R1` *contains* `[8 125 729]` — *Each element of* `V1` *is cubed*

```
R2=2.^V3;
```

`R2` *contains* `[16 0.0078 1]` — *The number 2 is raised to each element of* `V3`

```
R3=V3.^V1;
```

`R3` *contains* `[16 -16807 0]` — *Raise each element of* `V3` *to the corresponding element of* `V1`.

```
R4=V2.^V1;
```

Will return the message:

```
Error using .^ Matrix dimensions must agree.
```

Combined Operations

All of the standard rules concerning priority of operators, use of parentheses, etc. apply to computations with vectors, the only difference being that both intermediate and final results may be vectors instead of scalars.

EXAMPLE 16-8

What is stored in the variable when each line of MATLAB code below is exectuted?

Assume `V4=[2 1 -2]; V5=[3 -2 1.5];`

```
R=(10-V4./V5).^2*3;
```

`R` *contains* `[261.33 330.75 385.33]`

Sequence of operations:

1. `V4` *is divided element-wise by* `V5` — `[0.6667 -0.5 -1.3333]`
2. *Each of these values is subtracted from 10* — `[9.3333 10.5 11.3333]`
3. *Each of those values is squared* — `[87.1111 110.25 128.4444]`
4. *Each of those values is multiplied by 3* — `[261.33 330.75 385.33]`

COMPREHENSION CHECK 16-6

Assume a row vector named `Vals` has already been defined. Write a single MATLAB statement that will perform the following calculation using each element of `Vals` and leave the results in a vector named `Comps`: $C = (3V + 5)^4 - 16$ where V represents the individual values in `Vals`, and C represents the individual results in `Comps`. Example: if `Vals=[2 0 -1]`, `Comps=[14625 609 0]`

Functions Used with Vectors

Most of the functions listed in Table 16-4 also work with vectors. In most cases, such as `sqrt`, `sin`, or `round`, these automatically perform the stated operation on each element individually. There are quite a few MATLAB functions that would seldom if ever be used with scalars, but are very helpful in dealing with vectors. A few of these are listed in Table 16-5. You can obtain more information by typing either `help` or `doc` followed by the function name.

All of the functions in Table 16-5 are actually matrix functions. Matrices will be covered in the next section.

The last five functions, as well as functions like `min` and `max`, can also find the locations of the desired values in the vector.

Table 16-5 Selected Functions Used with Vectors

Function	Description
`min`	Find smallest value. Can also find location of smallest value in the vector.
`max`	Find largest value. Can also find location of largest value in the vector.
`sort`	Sort elements in ascending or descending order.
`zeros`	Create a vector of all zeros.
`ones`	Create a vector of all ones.
`rand`	Create a vector of random numbers.
`length`	Find the number of elements in a vector. VERY useful!
`unique`	List all values in a vector with no repetitions in result.
`intersect`	Find all values in both of two vectors, no repetitions in result
`union`	Find all values in either of two vectors, no repetitions in result
`setdiff`	Find all values in one vector that are not in another vector, no repetitions in result.
`setxor`	Find all values in each of two vectors that are not in the other, no repetitions.

Other Notes on Creating Vectors

You can include calculations, functions, etc. inside the brackets when defining vectors.

Examples:

NOTE

To **concatenate** means to combine

1. Vectors can be combined (concatenated) by simply listing them in the desired order inside of square brackets. For row vectors, the individual items should be separated by spaces or commas; for column vectors they should be separated by semicolons. Use the single quote as needed to transpose a row to column before combining.

```
V4=[2 1 -2];    V6=[3;-2;1.5];
CV=[V4,V6'];
```

CV contains [2 1 −2 3 −2 1.5]

2. You wish to create a two element vector named `MM` containing the maximum and minimum values in vector `V100`. This can be done using

```
MM=[max(V100), min(V100)];
```

3. You wish to create a vector `OddMM` consisting of the odd integers from 1 to 50 followed by three times the square root of each element in `MM` from the previous example.

```
OddMM=[1:2:50, 3*sqrt(MM)];
```

COMPREHENSION CHECK 16-7

(a) Assume you have four row vectors containing data on traffic flow named `T1`, `T2`, `T3`, and `T4`. Using a single MATLAB statement, combine these four, in that order, into a single column vector named `TFC`. Example:

```
T1=[2 6 4]; T2=[0 -1]; T3=[9 -1 0];
T4=[7 7]; TFC=[2;6;4;0;-1;9;-1;0;7;7]
```

(b) Using a single MATLAB statement, create a row vector named `Rev` that contains all even integers from 10 to 10,000 in ascending order followed by all integer multiples of 7 from 700 to 7 in descending order.

(c) Assume you have a row vector named `RV5`. Using a single MATLAB statement, create a new row vector named `Pow` that contains, in this order,

1. The elements of `RV5`.
2. The square roots of each element of `RV5`.
3. The square of each element of `RV5`.

Vector Indexing

IMPORTANT

Index values MUST be integers and strictly positive. Talking about the 2.7th element or the −3rd value does not make sense.

Frequently you will need to change a single element or group of elements within a vector, or perform some operation on only some of the elements within a vector. **Indexing** or **addressing** allows you to accomplish this. To access a specific element or elements within a vector, we place the numeric positions of the elements we wish to use in parentheses after the vector name. For example, in the vector `V1 = [2 5 9]`, the statement `V1(2)` indicated the second element in `V1`, or the value 5.

Creating Vectors Using Indexing

When creating a previously undefined vector using indexing, MATLAB will, by default, make it a row vector. If you need a column vector instead, you can transpose the vector. We will soon see how to create a column vector directly with indexing.

We have already seen that we can place a list of values in brackets to create a vector. We can also create vectors using indexing. As an example, assume the vector `AVec` containing the values 2, 0, and 5 could be created by

```
AVec=[2 0 5];
```

This could also be accomplished using

```
AVec(1)=2; % Define Element 1
AVec(2)=0; % Define Element 2
AVec(3)=5; % Define Element 3
```

Although this may look like a lot more work at first, as we will see later, this is the best way to define vectors in some situations.

What happens if you wish to specify some element of an undefined vector other than the first element?

In the command window, type this to be sure try3 does not already exist:

```
clear try3
```

Now type

```
try3(3)=9
```

Do not place a semicolon at the end of the line so that `try3` will be echoed to the screen. What can you say about the first two elements, `try3(1)` and `try3(2)`, that were not defined?

Changing Individual Values of Vectors

You can change elements individually without modifying the others using vector indexing.

EXAMPLE 16-9

What is stored in the variable when each line of MATLAB code below is exectuted?

Note that vector indexing works fine with column vectors. It is just the creation of new vectors where row is the default format.

Assume V7 = $\begin{bmatrix} 3 \\ -7 \\ 34 \\ 0 \\ 18 \end{bmatrix}$

```
V7(3) = -1;
```

V7 will then contain $\begin{bmatrix} 3 \\ -7 \\ -1 \\ 0 \\ 18 \end{bmatrix}$

Deleting Values from Vectors

Sometimes you need to simply delete elements from a vector—not replace them with zeros, but actually delete them making the vector shorter. This can be done by specifying the elements to be deleted to be the **empty element**, indicated by [].

EXAMPLE 16-10

What is stored in the variable when each line of MATLAB code below is exectuted?

Assume V8 contains [2 1 5 8 6 3].

```
V8(3)=[];
```

This will delete the third element, so V8 now contains [2 1 8 6 3].

Specifying More Than One Element Using Indexing

If you wish two or more elements of a vector to have the same value, you can list the indexes of those values in the parentheses separated by commas.

EXAMPLE 16-11

What is stored in the variable when each line of MATLAB code below is exectuted?

Assume V9 contains [3 −2 5 0 −7].

```
V9(2,4,5)=99;
```

This will replace the 2nd, 4th and 5th elements with the number 99, so V9 now contains [3 99 5 99 99].

Using Linear Sequences as Indexes

You can use the colon operator to create a linear sequence of values to use as indices inside the parentheses.

EXAMPLE 16-12

What is stored in the variable when each line of MATLAB code below is exectuted?

Assume CV1 = $\begin{bmatrix} 3 \\ -7 \\ 34 \\ 0 \\ 18 \end{bmatrix}$

```
CV1(2:4)=99;
```

This will replace the 2nd, 3rd and 4th elements with the number 99, so CV1 *now contains*

$$\begin{bmatrix} 3 \\ 99 \\ 99 \\ 99 \\ 18 \end{bmatrix}$$

EXAMPLE 16-13

What is stored in the variable when each line of MATLAB code below is exectuted?

Assume RV1 contains [3 6 −2 7 0 0 −4 −6 9 13].

```
RV1(2:3:10)=0;
```

This will replace the every third element, starting with the 2nd element (replacing the 2nd, 5th and 8th . . .) up to the 10th element, so RV1 *now contains [3 0 −2 7 0 0 −4 0 9 13].*

Calculations with Part of a Vector

EXAMPLE 16-14

Determine the sine of the fifth element of vector T20, assuming the values in T20 are angles given in radians. Place the result in S5.

```
S5=sin(T20(5));
```

EXAMPLE 16-15

Assume you have a 100-element vector D1 containing data. Create a new vector RD1 containing 50 elements comprising the square roots of the even numbered elements of D1.

```
RD1=sqrt(D1(2:2:100));
```

EXAMPLE 16-16

Assume you have a 1000-element vector D2 containing data. Create a new vector ZRD2 containing 1000 elements in which the odd numbered elements are zero and the even numbered elements are the squares of the corresponding even numbered elements of D2.

```
ZRD2(2:2:1000)= D2(2:2:1000).^2;
```

EXAMPLE 16-17

Assume you have a vector D3 containing an unknown number of elements. Modify D3 so that all odd numbered elements are replaced by the cosines of the odd numbered values. All even numbered values should be unchanged. Assume the odd numbered elements contain angles in degrees.

```
D3(1:2:length(D3))= cosd(D3(1:2:length(D3)));
```

WARNING

If you try to do a calculation on a vector element that does not exist, such as the sixth element of a five element vector, you will get an error. The `length` function can prove very useful if you do not know how many elements a vector has, or if the number of elements might change during execution of the program.

COMPREHENSION CHECK 16-8

(a) Write a single MATLAB command that will create a 50,000-element vector named `Biggie` in which every tenth element equals −9999 and all other elements equal 0.

(b) Assume a vector `LV` has already been defined. Write a single MATLAB command that will delete all of the even numbered elements, leaving `LV` with half as many elements.

(c) Assume a vector named `D` has an even number of elements. Write a single MATLAB command that will create a new vector named `DS` with half as many elements, in which each element is the sum of adjacent odd-even pairs in `D`.

Example:

If `D= [2 5 4 -7 3 0]`, then `DS` will equal `[7 -3 3]`, found by the vector:

$[D(1)+D(2) \quad D(3)+D(4) \quad D(5)+D(6)]$

16.4 MATRICES

LEARN TO:
Define numeric matrices in MATLAB's workspace
Perform basic matrix operations
Write mathematic expressions using functions on an entire matrix

A **matrix** or **array** consists of numbers organized into rows and columns. The size of a matrix is given as `R × C`, where R is the number of rows and C is the number of columns. Examples include $\begin{bmatrix} 0.2 & 3 & 0 \\ -7 & 0.05 & 99 \end{bmatrix}$ which is a 2×3 matrix, and $\begin{bmatrix} 1 & 0 & 0 & 0 \\ 1 & 1 & 0 & 1 \\ 1 & 0 & 1 & 0 \\ 0 & 1 & 1 & 1 \end{bmatrix}$ which is a 4×4 matrix. All numeric data in MATLAB, including scalars and vectors, are matrices. MATLAB is, after all, derived from the words **MAT**rix **LAB**oratory. A scalar is simply a 1×1 matrix, whereas a vector is either $1 \times C$ (row) or $R \times 1$ (column). Note that all numeric data, whether scalar, vector, or matrix, is represented in the workspace by a square icon divided into four smaller squares used to symbolize a matrix.

Defining Matrices

The basic way to define a matrix is to combine the methods used to create row and column vectors, but matrices are created row by row, not column by column—you specify the first row with elements separated by commas or spaces, then use a semicolon to move to the next row.

EXAMPLE 16-18

What matrix is formed by the following commands?

```
M1=[0 -4 3;-2 6 0.2];
```

Creates the 2 × 3 matrix M1 $= \begin{bmatrix} 0 & -4 & 3 \\ -2 & 6 & 0.2 \end{bmatrix}$

```
M2=[0,-4;3,-2;6 0.2];
```

Creates the 3 × 2 matrix M2 $= \begin{bmatrix} 0 & -4 \\ 3 & -2 \\ 6 & 0.2 \end{bmatrix}$

Creating Matrices Using Indexing

When using indexing with matrices, we need to specify both the row and column of the element in question. Instead of a single number, place two integers separated by a comma in the parentheses following the matrix name. The first value is the row number, and the second value is the column number.

EXAMPLE 16-19

What matrix is formed by the following commands?

```
M3(1,1)=5;
M3(2,1)=0;
M3(1,2)=1;
M3(2,2)=8;
```

Creates the 2 × 2 matrix M3 $= \begin{bmatrix} 5 & 1 \\ 0 & 8 \end{bmatrix}$

Note that when using indexing to create a matrix, the order in which the matrix is defined is not predefined. In the M3 example, the matrix was defined column by column, not row by row.

If you create an element of a matrix in such a way that for the row and column specified there are elements of the matrix that are unspecified, MATLAB will automatically make them zero.

EXAMPLE 16-20

What matrix is formed by the following commands?

```
M4(2,3) = 77;
```

Creates the 2 × 3 matrix M4 $= \begin{bmatrix} 0 & 0 & 0 \\ 0 & 0 & 77 \end{bmatrix}$

```
M5(2,3)=55;
M5(3,2)=88;
```

Creates the 3 × 3 matrix M5 $= \begin{bmatrix} 0 & 0 & 0 \\ 0 & 0 & 55 \\ 0 & 88 & 0 \end{bmatrix}$

Using the Colon Operator to Create Linear Sequences for Indexing Matrices

EXAMPLE 16-21 What matrix is formed by the following commands?

Assume M5 has already been defined as in the previous example.

```
M5(1,2:3) = -1;
```

This command states that Row 1, Columns 2 through 3 are set to −1.

M5 *now contains* M5 $= \begin{bmatrix} 0 & -1 & -1 \\ 0 & 0 & 55 \\ 0 & 88 & 0 \end{bmatrix}$

Defining a Matrix from Another Matrix

EXAMPLE 16-22 What matrix is formed by the following commands?

Assume M5 has been defined as in the previous example.

```
M6=M5(1:2,2:3);
```

This command states the variable M6 *equals Rows 1 and 2, Columns 2 and 3 of* M5.

This creates M6 $= \begin{bmatrix} -1 & -1 \\ 0 & 55 \end{bmatrix}$

NOTE

A **colon** used as an index indicated to use all rows or all columns in the matrix.

Accessing Entire Rows or Columns

Often you wish to specify an entire row or entire column or a matrix. MATLAB provides a shortcut notation for this situation. Instead of placing a 1 for the first row (or column) before the colon and the total number of rows (or columns) after the colon, simply type a colon by itself.

EXAMPLE 16-23 What matrix is formed by the following commands?

Assume M5 has been defined as above.

```
M7=M5(:,2);
```

This command states the variable M7 *equals all rows, Column 2 of* M5.

This creates the 3 × 1 matrix (or column vector) M7 $= \begin{bmatrix} -1 \\ 0 \\ 88 \end{bmatrix}$

```
M8=M5(1:2:3,:);
```

This command states the variable M8 *equals Rows 1 and 3, all columns of* M5.

This creates the 2 × 3 matrix M8 $= \begin{bmatrix} 0 & -1 & -1 \\ 0 & 88 & 0 \end{bmatrix}$

Creating a Column Vector (or a R × 1 Matrix)

You can use the colon to force a vector definition to be a column instead of the default row.

EXAMPLE 16-24

What matrix is formed by the following commands?

```
CV2(:,1)=[1 4 3 7];
```

This command states the variable `CV2` *equals all rows, one column of values. This creates*

$$\text{CV2} = \begin{bmatrix} 1 \\ 4 \\ 3 \\ 7 \end{bmatrix}$$

Deleting Rows or Columns from a Matrix

The colon operator in combination with the empty element [] can be used to delete entire rows or columns from a matrix.

EXAMPLE 16-25

What matrix is formed by the following commands?

$$\text{Assume M9} = \begin{bmatrix} 3 & -9 & 14 \\ 6 & 0.5 & -4 \\ 44 & 5 & 1 \end{bmatrix}$$

```
M9(2,:)=[];
```

This command replaces Row 2, all columns with an empty matrix, deleting Row 2 from the matrix. Now,

$$\text{M9} = \begin{bmatrix} 3 & -9 & 14 \\ 44 & 5 & 1 \end{bmatrix}$$

Deleting Individual Elements from a Matrix

EXAMPLE 16-26

What matrix is formed by the following commands?

$$\text{Assume M10} = \begin{bmatrix} -1 & -2 \\ 3 & 5 \end{bmatrix}$$

```
M10(2,1)=[];
```

This command replaces Row 2, the first column with an empty matrix. This will result in the error:

```
Subscripted assignment dimension mismatch.
```

The error occurs because in a matrix, all rows must have the same number of columns and all columns must have the same number of rows. Deleting a single element from a row or column would violate this rule. You can, however, replace elements with zeros.

```
M10(2,1)=0;
```

This command replaces the element in Row 2, Column 1 with a zero, so M10 $= \begin{bmatrix} -1 & -2 \\ 0 & 5 \end{bmatrix}$

Matrix Transpose

In the section on vectors, we saw that the single quote (') operator changes a row vector to a column vector, or vice versa. This is actually just a special case of a matrix transpose. The transpose of a matrix swaps rows and columns—the first row of the original matrix becomes the first column of the transposed matrix, the second row becomes the second column, etc.

EXAMPLE 16-27

What matrix is formed by the following commands?

Assume M1 $= \begin{bmatrix} 0 & -4 & 3 \\ -2 & 6 & 0.2 \end{bmatrix}$

```
T1 = M1';
```

This will transpose the M1 *matrix, and store it in* T1: T1 $= \begin{bmatrix} 0 & -2 \\ -4 & 6 \\ 3 & 0.2 \end{bmatrix}$

COMPREHENSION CHECK 16-9

(a) Create the matrix CCM1 $= \begin{bmatrix} 18 & 0.3 \\ -4.1 & -1 \\ 0 & 17 \end{bmatrix}$ using a single MATLAB command.

(b) Create the matrix CCM2 $= \begin{bmatrix} 0 & 0 & 0 \\ 0 & 0 & 1 \times 10^{15} \\ 0 & 0 & 1 \times 10^{15} \end{bmatrix}$ using a single MATLAB command.

(c) Assume the 3×3 matrix CCM3 has already been defined as CCM3 $= \begin{bmatrix} 3 & 9 & 14 \\ 6 & 0.5 & 4 \times 10^{-2} \\ 44 & 0 & 1 \times 10^{3} \end{bmatrix}$

Using a single MATLAB command, define a new 2×2 matrix named Corners that contains the corner elements of CCM3. Your code should still work correctly if the contents of CCM3 changes but remains a 3×3; in other words, DO NOT hard-code the four values.

Calculations with Matrices

For the basics of matrix operations, please refer to Appendix A.6 online. The following discussion assumes that you already understand things like a matrix transpose or matrix multiplication.

Addition and Subtraction

When addition or subtraction involves a matrix, there are two cases:

1. A scalar and a matrix: the scalar is added or subtracted to each element of the matrix.
2. Two equal size matrices: corresponding elements of the two matrices are added or subtracted. If you attempt to add two matrices (both non-scalar) that are not the same size, you will get angry red letters.

EXAMPLE 16-28 What matrix is formed by the following commands?

Assume `ExM1` $= \begin{bmatrix} 3 & 0.5 & 0 \\ 5 & -7 & 2 \end{bmatrix}$; `ExM2` $= \begin{bmatrix} 9 & -4 \\ 8 & 3.7 \end{bmatrix}$; `ExM3` $= \begin{bmatrix} 19 & 2 & 3.4 \\ -6 & 1 & -3 \end{bmatrix}$;

```
D1=7-ExM1;
```

`D1` *contains* $\begin{bmatrix} 4 & 6.5 & 7 \\ 2 & 14 & 5 \end{bmatrix}$

```
S1=ExM1 + ExM3;
```

`S1` *contains* $\begin{bmatrix} 22 & 2.5 & 3.4 \\ -1 & -6 & -1 \end{bmatrix}$

```
E1=ExM1 + ExM2;
```

Will return the message:

```
Error using + Matrix dimensions must agree.
```

Multiplication and Powers

We will consider four cases of matrices involved in multiplication and powers.

1. A scalar and a matrix: The scalar is multiplied by each element of the matrix.
2. Two equal size matrices multiplied on an element-by-element basis. Note that the two matrices MUST have the same dimensions. NOTE: This requires the use of the element-wise operator as shown below.
3. A true matrix multiplication (refer to Appendix A.6 online): In this case, the number of columns of the first matrix MUST equal the number of rows of the second matrix or the matrix multiplication is not defined.
4. A matrix raised to a scalar. Until you understand considerably more about matrix arithmetic, it would be wise to avoid fractional or negative powers when using ^ (full matrix power). Negative and fractional powers are fine with element-by-element powers (.^)—they work just as you would expect.

EXAMPLE 16-29 What matrix is formed by the following commands?

Assume ExM1 $= \begin{bmatrix} 3 & 0.5 & 0 \\ 5 & -7 & 2 \end{bmatrix}$; ExM2 $= \begin{bmatrix} 9 & -4 \\ 8 & 3.7 \end{bmatrix}$; ExM3 $= \begin{bmatrix} 19 & 2 & 3.4 \\ -6 & 1 & -3 \end{bmatrix}$

RV = [2 5 9]; CV $= \begin{bmatrix} 3 \\ -1 \\ 5 \end{bmatrix}$

```
P1=3*ExM1;
```

P1 *contains* $\begin{bmatrix} 9 & 1.5 & 0 \\ 15 & -21 & 6 \end{bmatrix}$

```
P2=ExM1.*ExM3;
```

P2 *contains* $\begin{bmatrix} 57 & 1 & 0 \\ -30 & -7 & -6 \end{bmatrix}$ *Recall the symbol .* means element-wise multiplication*

```
P3=ExM1*ExM3;
Error using * Inner matrix dimensions must agree.
P4=RV*CV;
```

P4 *contains 46*

```
P5=CV*RV;
```

P5 *contains* $\begin{bmatrix} 6 & 15 & 27 \\ -2 & -5 & -9 \\ 10 & 25 & 45 \end{bmatrix}$

P4 *and* P5 *show that matrix multiplication is NOT commutative!*

```
P6=ExM2*ExM1;
```

P6 *contains* $\begin{bmatrix} 7 & 32.5 & -8 \\ 42.5 & -21.9 & 7.4 \end{bmatrix}$

```
P7=ExM1*ExM2;
Error using * Inner matrix dimensions must agree.
```

P6 *and* P7 *show that commuted matrix multiplication is not necessarily even defined!*

```
P8=ExM2^2;
```

P8 *contains* $\begin{bmatrix} 49 & -50.8 \\ 101.6 & -18.31 \end{bmatrix}$

```
P9=ExM1^2;
Error using ^ Inputs must be a scalar and a square matrix.
To compute elementwise POWER, use POWER (.^) instead.
P10=ExM1.^2;
```

P10 *contains* $\begin{bmatrix} 9 & 0.25 & 0 \\ 25 & 49 & 4 \end{bmatrix}$ *Recall using symbol .^ means element-wise power*

Division

The only case we will consider in this text is division of a matrix by a scalar. In this case each element of the matrix is divided by the scalar. Division by a matrix is beyond the scope of this text. Refer to any general reference on matrix algebra for more information.

EXAMPLE 16-30 What matrix is formed by the following commands?

Assume `ExM2` $= \begin{bmatrix} 9 & -4 \\ 8 & 3.7 \end{bmatrix}$

```
Q1=ExM2/4;
```

`Q1` *contains* $\begin{bmatrix} 2.25 & -1 \\ 2 & 0.925 \end{bmatrix}$

Functions Used with Matrices

NOTE

All of the functions used with vectors in Table 16-5 can be used with matrices, but some do NOT do what you might expect. For example, `length` does **not** find the total number of elements in a matrix and `max` does **not** find the single largest element in the matrix.

Most of the functions listed in Table 16-4 and 16-5 also work with matrices. In most cases, such as `sqrt`, `sin`, or `round`, these automatically perform the stated operation on each element individually. There are quite a few MATLAB functions that would seldom if ever be used with scalars and/or vectors, but are very helpful in dealing with matrices. A few of these are listed in Table 16-6. You can obtain more information by typing either `help` or `doc` followed by the function name.

Table 16-6 Selected Functions Used with Matrices

Function	Description
`size`	Find the number of rows and columns in a matrix. VERY useful!
`eye`	Create identity matrix with R rows and C columns, where all elements are zero except the diagonal
`inv`	Find the inverse of a matrix.
`reshape`	Create matrix containing the same elements but different dimensions
`ndims`	Determines the number of dimensions of a matrix
`diag`	Specify or extract diagonals of matrix
`trace`	Sum of elements on the diagonal
`flipud`	Flip matrix in up/down direction
`fliplr`	Flip matrix in left/right direction
`rot90`	Rotate matrix 90 degrees counterclockwise
`sparse`	Create sparse matrix; removes all zero elements from a matrix
`find`	Find indices of all non-zero elements in a matrix
`full`	Convert sparse matrix to full matrix

EXAMPLE 16-31 What matrix is formed by the following commands?

Assume ExM1 $= \begin{bmatrix} 3 & 0.5 & 0 \\ 5 & -7 & 2 \end{bmatrix}$

```
[R,C]=size(ExM1);
```

R *contains 2 and* C *contains 3*

```
Ident=eye(3);
```

Ident *contains* $\begin{bmatrix} 1 & 0 & 0 \\ 0 & 1 & 0 \\ 0 & 0 & 1 \end{bmatrix}$

```
Mod1=reshape(ExM1,3,2);
```

Mod1 *contains* $\begin{bmatrix} 3 & -7 \\ 5 & 0 \\ 0.5 & 2 \end{bmatrix}$

```
Mod2=fliplr(ExM1);
```

Mod2 *contains* $\begin{bmatrix} 0 & 0.5 & 3 \\ 2 & -7 & 5 \end{bmatrix}$

COMPREHENSION CHECK 16-10

(a) Create a matrix CC1 defined as the square root of half of each element in matrix QPD.

(b) Create a matrix CC2 defined as 17 plus the sum of corresponding elements in matrices X1 and X2. Assume X1 and X2 are the same size.

16.5 CHARACTER STRINGS

LEARN TO:
Define text strings in MATLAB's workspace
Combine existing text strings
Write expressions using special functions to modify character strings

Character Strings abc

A variety of data, such as names, part identification, or week days, requires alphabetic or alphanumeric data. Such information is stored in a **character string**, also called a text string or simply text, and is denoted in the workspace with a square icon containing the letters abc.

A character string is in one sense like a numeric vector, in that each character of the string is in a separate element of the variable where it is stored. For example, if the variable Z contains the text string 'Cat', the second element of Z contains the character 'a'. Examples of text strings include 'Ignatius J. Reilly', 'Part # 23B', and 'Wednesday'.

Note that punctuation marks and spaces each require a separate element within the variable. For example, `'Part # 23B'` is a ten element character string; elements 5 and 7 are the blank character, and element 6 is the number symbol #. Also, numerals in a text string each require a separate element, and are encoded in binary very differently from the corresponding numeric value; in this example, elements 8 and 9 are the numerals 2 and 3 encoded as characters.

TEXT CREATION

We saw earlier that the single quote is used as the transpose operator. Single quotes are also used to define character strings. MATLAB is clever enough to understand the meaning of the quote from context: if a single quote immediately follows a matrix, it is the transpose operator; if two single quotes surround characters, they are being used to indicate that those characters are simply text, not variables or calculations or MATLAB commands.

Defining Character Strings

When creating a character string using the assignment operator (=), simply enter the desired text enclosed in single quotes.

```
TS1='American Bison';
TS2='Tuesday, May 11, 868';
TS3='42';
```

Note that `TS3` is a text string, not a numeric value, and consists of two characters, `'4'` and `'2'`.

Including an Apostrophe in a Text String

If you wish to create a text string including an apostrophe, there is a minor problem since MATLAB will think the apostrophe is the end of the text string, not a character to be included in the text string. However, MATLAB also knows that if it is processing a text string and finds TWO single quotes in a row (2 single quotes, NOT one double quotation mark) it knows to place a single apostrophe in the text string, not terminate the string. Thus, if you want to place the text "Willy's brain" into variable `WB`, you would type `WB='Willy''s brain';`

Combining Text Strings

Just like you can combine vectors by concatenation in square brackets, you can combine text strings.

```
W1='Willy''s';
W2='Brain';
TS=[W1,' Weird ',W2];
```

`TS` contains `Willy's Weird Brain`. Note the spaces in the single quotes before and after `Weird`.

Using Indexing with Character Strings

Character strings can be created or modified using indexing, and indexing can also be used to extract portions of a text string.

```
TS='Willy''s Weird Brain';
Word1=TS(1:7)
```

`Word1` contains `Willy's`. Note the apostrophe is ONE character.

```
W2=TS(length(TS)-3:length(TS));
```

`Word2` contains `rain`.

```
TS(9:14)=[];
```

`TS` now contains `Willy's Brain`

Functions Used with Character Strings

Some of these functions (e.g., `isletter`) return a logical array. Logical arrays will be covered in a later chapter.

Although many of the standard mathematical operators and functions will actually process character strings without errors, you may not get the results you expect, and you need to be certain that you understand what actually happens before trying to use these with text.

There are, however, numerous functions specifically for manipulating text, several of which are included in Table 16-7 below.

Table 16-7 Selected Functions Used with Character Strings

Function	Description
`blanks`	Create a string of blanks (compare the zeros function used with matrices)
`char`	Convert ASCII codes into characters
`ischar`	Determines if a variable is a character string
`strcat`	Concatenate strings
`isletter`	Finds alphabetic characters in a string
`isspace`	Finds space characters in string
`isstrprop`	Finds elements in string meeting certain criteria, such as lowercase or punctuation
`strfind`	Find a string within another string
`strrep`	Replace substring within a string
`strcmp`	Compare two strings
`strcmpi`	Compare two strings ignoring case
`lower`	Convert string to all lowercase
`upper`	Convert string to all uppercase
`num2str`	Convert number to string
`str2num`	Convert string to number

COMPREHENSION CHECK 16-11

(a) Create a variable named `MTS` containing the text: `My hero's hat`

(b) Create a variable named `B` containing 17 blanks followed by your name followed by 691 more blanks.

(c) Assume `Avian='A wet bird never flies at night';`
Modify `Avian` so that the word "wet" is replaced with the word "crepuscular" and the word "never" is replaced by "seldom". Note that you should actually modify ONLY the two words in question, not simply type in the complete new sentence.

(d) Assume `Ultimate='The answer is 42';`
Convert the substring "42" in `Ultimate` into its corresponding numeric value and place it in `UltAns`. Your solution should work as long as the "number" at the end of the string is exactly two digits.

16.6 CELL ARRAYS {}

LEARN TO: Define cell arrays in MATLAB's workspace
Perform basic operations on cell arrays
Write expressions using special functions to modify cell arrays

So far in our discussion of data types in MATLAB, all elements stored in a specific variable are the same type of thing: either all numeric or all text. Sometimes, however, it would be convenient to be able to store different types of information in a single variable. MATLAB provides two data types that allow us to store both numeric values and non-numeric values such as text within a single structure stored in a variable. The first of these structures is called a **cell array** or **cell matrix**.

Cell Arrays

The basic idea of the cell array is that each element in a cell array contains another MATLAB data structure. An element in a cell array might be a numeric matrix, whether a scalar, a vector or a matrix; it might be a text string; or it might even be another cell array.

Defining Cell Arrays

Recall that when we define numeric variables, scalars are defined using a simple assignment statement such as `S1 = 57.43`, whereas vectors and matrices are defined by enclosing the multiple values in square brackets, e.g., `M1 = [3.5, − 9;13, − 4.2]`. Text strings are defined using single quotes: `TS = 'ARF'`. We also saw that we could define individual elements of matrices or text strings using indexing. For example `M2(2,4) = 42` places the value 42 into the second row, fourth column element of `M2` and `TS1(5:9)='BOOF!'` places the text string `BOOF!` into elements 5 through 9 of `TS1`.

Cell arrays are defined in a similar manner, with one key difference: the use of curly braces { }.

EXAMPLE 16-32 What is stored in the variable by following commands?

Assume `M3=[3, 5; 1 9]` and `TS3 = 'GRRR'`
`CA1={TS3,42,M3,'RECWR'}`
`CA1` *now contains four elements in a single row:*
`CA1{1}` *The text string* `GRRR`
`CA1{2}` *The number* `42`
`CA1{3}` *A 2x2 numeric matrix with the same elements as* `M3`
`CA1{4}` *The text string* `REOWR`
Note that the individual elements are different types of things: a scalar, a matrix, and two text strings.
Note the use of curly braces to index into the cell array.

We could also use indexing to define one or more elements of a cell array separately instead of all at once. The cell array CA1 above could also be created with the following code:

```
CA1(1) = {'GRRR'}; % braces tell MATLAB CA1 is a cell array
CA1(2:2:4) = {42,'REOWR'}; % CA1(3) will be created empty
CA1(3) = {[3,5;1,9]}; % Go back to specify CA1(3)
```

Note that the cell array indices are in parentheses, whereas the contents to be placed in the indexed values are in curly braces. This might be confusing at first since we used parentheses here and referred to using curly braces to indexing cell arrays above. We will clarify the difference in the use of the two notations in the next section.

COMPREHENSION CHECK 16-12

(a) Create a cell array named Cabinet with one row and three columns. The first cell should contain the number 77, the second cell should contain the text "Dr. Caligari", and the third cell should contain a 25 × 100 matrix filled with ones.

(b) Create a 2 × 2 cell array named Cyls containing the following:

a. First row: "Diameter", "Length"

b. Second row: 1.5, a column vector containing all integers from 1 to 25

Extracting Data from Cell Arrays

When we use indexing with parentheses to get information out of a numeric matrix, the result is another numeric matrix. If M4=[1 3 5;6 8 0] and we create another variable as M5=M4(:,2:3), then M5 is a 2 × 2 numeric matrix containing [3 5; 8 0].

When we use indexing with parentheses to get information out of a text string, the result is another text string. If TS4='Dogs vs Cats' and we create another variable as TS5=TS4(4:9), then TS5 is a text string containing s vs C.

In general, when we access some of the elements of a variable using index values in **parentheses**, the result is the same **variable type** as the original variable.

But what about cell arrays, since they often contain different variable types in the various elements?

This is perhaps best illustrated live at the computer, so we suggest that you fire up MATLAB and type the following simple commands in the Command Window.

First, we will set up three variables. Do not include a semicolon so each echoes to the screen.

Type:

```
S6=777 % define a scalar
```

The following appears on the screen:

```
S6=
   777
```

Next, type:

```
M6=[9 7 8; 3 5 4] % define a 2 × 3 matrix
```

The following appears on the screen:

```
M6=
   9  7  8
   3  5  4
```

Finally, type:

```
TS6='apricot' % Define a text string
```

The following appears on the screen:

```
TS6 =
apricot
```

As you will see in a moment, it is important that you note exactly how these values are echoed to the screen.

Next, we will create a cell array containing the scalar, the matrix, and the text string created above. Add a semicolon to suppress the output in this case, however.

```
CA6={S6,M6,TS6}; % Create cell array with three elements
```

If you look at `CA6` in the workspace, it will say <1×3 cell>.

Now we will attempt to extract each of these three elements from the cell array using parentheses to get the original contents back. First, extract the scalar in element 1.

```
CA7=CA6(1) % Extract first element
```

Echoed to the screen is

```
CA7 =
      [777]
```

Comparing this to the echoed value from the original scalar `S6` above, we see there is a slight difference: there are brackets around the value!

Now do the same with the matrix in the second element of the cell array.

```
CA8=CA6(2) % Extract second element
```

Echoed to the screen is

```
CA8 =
      [2×3 double]
```

This is even stranger. The original matrix echoed the individual values to the screen, but when we attempted to extract it from the cell array, it just gave us the matrix dimensions in brackets.

Finally, extract the third element, the text string.

```
CA9=CA6(3) % Extract third element
```

The echo to the screen is

```
CA9=
   'apricot'
```

Comparing this to the echo of the original text string, we see that the text is enclosed in single quotes, whereas in the original echo of `TS6`, there were no quotes.

Obviously something is different here, and in fact, if you were to try to use these extracted values functions or computations that expected a scalar, a matrix, or a text string respectively, you would get an error message. For example, type

```
SUM=CA7+111
```

This seems perfectly reasonable, and should give an answer of 888. However, you get the error message

```
Undefined function 'plus' for input arguments of type 'cell'.
```

The point is that when parentheses are used to extract values from a numeric matrix, the result is a numeric matrix, from a text string the result is a text string, and **values extracted from a cell array are a cell array**, thus they cannot be used where MATLAB is expecting numbers or text.

Obviously there must be a way to overcome this problem or cell arrays would be extremely limited in their utility. The answer is embodied in the well-known alliterative mnemonic: **Curly for Contents**.

With the cell array CA6 still in your workspace, try extracting the **contents** using curly braces instead of parentheses:

Type:

```
S7=CA6{1}  % Extract contents first element
```

Echo:

```
S7 =
   777
```

Type:

```
M7=CA6{2}  % Extract contents second element
```

Echo:

```
M7 =
       9     7     8
       3     5     4
```

Type:

```
TS7=CA6{3}  % Extract contents third element
```

Echo:

```
TS7 =
apricot
```

Note that the echoes to the screen are exactly the same as when we originally defined them. You may now use S7, M7, and TS7 anywhere you can use a scalar, a matrix, or a text string respectively.

We can also extract parts of these variables using indexing, just like we could with the original variables.

For example:

V=M7(2,3) places the value 4 into V
T=TS7(5:7) places the text string cot into T

What if we wanted to place the second column of the second element of the cell array CA6 into CV1?

We could use

```
CV1=CA6{2};
CV1=CV1(:,2);
```

or we can use a shortcut method, directly indexing into the extracted matrix:

```
CV1=CA6{2}(:,2);
```

Although all of the above examples have used a cell array with a single row thus we used a single index value in the curly braces, cell arrays can have 2 dimensions thus two index values.

Now let us look at a slightly more complicated example. Again it is recommended that you type this in on the computer.

We will set up a cell array `CA12` containing two other cell arrays.

```
CA10={[8 6 4;3 7 0],'Occurrence'};
CA11={[2 3 4;7 6 5],'Then'};
CA12={CA10,CA11};
```

What if we wanted to add the two 2 × 3 matrices found in the two cell arrays of `CA12`? Try

```
Result=CA12{1}(1) + CA12{2}(1);
```

You might think that this will pull out the first elements of each of the two cell arrays in `CA12` (the two 2 × 3 matrices), but instead, we get the error

```
Undefined function 'plus' for input arguments of type 'cell'.
```

The reason is that since `CA12{1}` refers to CA10, which is itself a cell array, when we use parentheses to index, we get another cell array, NOT the original matrix. In this case we have to extract the contents of the contents, so both indices must have curly braces.

```
Result=CA12{1}{1} + CA12{2}{1};
```

This will give us the correct sum of the two matrices.

A similar situation exists if we wanted to extract part of the text strings. To get characters 2, 3, and 4 from `Then`, we would use

```
T=CA12{2}{2}(2:4);
```

`T` now contains `hen`.

- The first `{2}` pulls out the contents of the second element in `CA12`: a cell array.
- The second `{2}` pulls out the contents of the second element of that inner cell array: `Then`.
- The final `(2:4)` pulls the 2nd through 4th elements from that text string, placing `hen` into `T`.

COMPREHENSION CHECK 16-13

Assume a cell array named `CA` has three cells in a single row.

- The first cell contains a name in the format `X.Y.Family`. For example: `M.V.Smith`.
- The second cell contains a 2 × 3 matrix. For example: `[1 3 5;9 6 3]`
- The third cell contains a 3 × 2 matrix. For example: `[4 7;-2 -6;0 1]`

(a) Extract the family name (Smith in the example above) from the first cell and place it in column one of a newly created second row of `CA`. Your solution should work for family names of any length. Note that the first six characters of the name in the first cell will always be the two initials, each followed by a dot and a space.

(b) Multiply the matrix in the second cell of row 1 by the matrix in the third cell of row 1 and place the result in the second row, second column of CA.

(c) Calculate the element-wise product of the matrix in row 1, column 2 by the transpose of the matrix in row 1, column 3. Place the result in the third column of the second row of CA.

Functions Used with Cell Arrays

Some helpful functions for cell arrays are listed in Table 16-8. You can obtain more information by typing either `help` or `doc` followed by the function name.

Table 16-8 Selected Functions Used with Cell Arrays

Function	Description
`iscell`	Determines if a variable is a cell array
`cell`	Create a cell array filled with empty matrices
`num2cell`	Converts a numeric matrix into a cell array
`mat2cell`	Split matrix into sub-matrices and store as cell array
`cell2mat`	Convert cell array to matrix. All cells must be same type: numeric or text
`cellfun`	Apply a function to a cell array
`celldisp`	Display contents of cell array in command window

EXAMPLE 16-33

What is stored in the variable by following commands?

$$M1 = \begin{bmatrix} 1 & 4 & -8 & 0 \\ 9 & 3 & 12 & 6 \\ 0 & 1 & -5 & 2 \end{bmatrix}$$

```
T1 = 'Zero'
S1 = 42
CA1 = {M1,T1,S1}
```

`IC1=iscell(CA1);`

IC1 contains 1 (true) since the variable CA1 is a cell array

`IC2=iscell(M1);`

IC1 contains 0 (false) since the variable M1 is a matrix, not a cell array

`CA2=cell(4);`

CA2 is a 4 × 4 cell array of empty matrices

`CA3=cell(2,5);`

CA3 is a 2 × 5 cell array of empty matrices

`CA4=mat2cell(M1, [2 1], [3 1])`

`iscell` returns a logical variable. Logical variables will be covered in a later chapter.

Unsuppressed, the following appears on screen

```
CA4 =
    [2×3 double]      [2×1 double]
    [1×3 double]      [          2]
```

The command mat2cell will split the matrix `M1` *into the following combinations, starting in the upper, left corner of the matrix:*

$$\begin{bmatrix} 1 & 4 & -8 \\ 9 & 3 & 12 \end{bmatrix}$$

$$\begin{bmatrix} 0 \\ 6 \end{bmatrix}$$

$$[0 \quad 1 \quad 5]$$

$$[2]$$

```
celldisp(CA4)
```

The following appears on screen

```
CA4{1,1} =
     1    4    -8
     9    3    12
CA4{2,1} =
     0    1    -5
CA4{1,2} =
     0
     6
CA4{2,2} =
     2
```

```
[rows,cols]=cellfun(@size,CA4)
```

The following appears on screen

```
rows =
     2    2
     1    1
cols =
     3    1
     3    1
```

`cellfun` uses a handle (sort of like a pointer) to the function desired. The **@** symbol before the function name creates the appropriate handle, telling MATLAB to use that function on the cell array.

NOTE: This gives us the sizes of the four matrices in `CA4`.

`CA4{1,1}` *is 2 × 1*
`CA4{1,2}` *is 2 × 1*
`CA4{2,1}` *is 1 × 3*
`CA4{2,2}` *is 1 × 1*

COMPREHENSION CHECK 16-14

NOTE: You may want to use help or doc to investigate the operation of some of the functions in Table 16-7 to solve these problems.

(a) Convert an M × N matrix `BigMat` to an M × N cell array `BigCell` consisting of the scalar values from each corresponding element of `BigMat`.

Example: If `BigMat` is defined by `BigMat = [2 6;5 8]`, your code would produce the same cell array as `BigCell = {[2] [6]; [5] [8]}`. Your solution must, of course, work for any matrix, whether or not you know its contents.

(b) Assume a cell array `VCell` contains only numeric vectors. Create a matrix `VCellMax` with the same dimensions as the cell array containing the maximum value of each vector in the corresponding location of the cell array.

Example:
If `VCell={[1 5 2] [-3 -5 -6 -1]}`, then `VCellMax=[5 -1]`

16.7 STRUCTURE ARRAYS

As we saw in the previous section, we can use cell arrays to store different types of things in one variable: matrices, text, other cell arrays, etc. To access the different elements of a cell array, we use standard indexing, usually with curly braces, to extract the content of a specific cell or cells.

Sometimes it would be convenient if we could refer to different cells or groups of cell using a descriptive name instead of a number. We could assign specific index values to named variables and use these variable names instead of the numbers they hold to access elements from a cell array. However, there is a more direct method of using names to access specific elements within a mixed-content variable. Such a variable is called a **structure array,** and it inherently uses names instead of numbers to access different parts of the structure.

Structure Arrays

A structure array consists of one or more **structures**, each containing the same **fields**. Each field contains a value or values relating to a specific aspect of the structure.

As an example, we might have a structure array named `MetalData` containing information about metals. `MetalData` might contain the fields `Type` (Steel, Brass, Zinc, etc.), `SpecificGravity`, `ThermalConductivity`, and `Resistivity`.

Defining Structure Arrays

First, a quick review of defining the other data types we have considered.

When defining numeric variables, scalars are defined using a simple assignment statement such as `S1=-699`, whereas vectors and matrices are defined by enclosing the multiple values in square brackets, such as `M1=[3.5, -9; 13,-4.2]`. Indexing can be used to define individual elements or groups of elements. For example, `M2(4,3:5)=[9 6 3]` places the three values 9, 6, and 3 into the third, fourth, and fifth columns of the fourth row of `M2`.

Text strings are defined using single quotes: `TS1='Neon'`. Individual elements or substrings of text variables can be defined using indexing. For example `TS2(6:10)= 'Argon'` places the text string `Argon` into elements 6 through 10 of `TS2`.

Cell arrays are defined in a similar manner, except curly braces are used to enclose the elements composing the cell array. These elements might themselves be scalars, matrices (using brackets), text (using single quotes), or even other cell arrays.

Structure arrays, on the other hand, use a somewhat different scheme, perhaps best explained by way of an example. Type the following lines into the command window:

It may appear that these variables violate the naming rules (no periods allowed), but in the case of structure arrays, the period is not part of a name, but a separator between two names: the structure name and the field name.

```
MetalData.Type='Zinc';
MetalData.SpecGrav=7.14;
MetalData.ThermCond=116;
MetalData.Resistivity=59E-9;
```

Now the structure array `MetalData` contains a single structure comprising 4 fields. We can add another similar structure to the `MetalData` structure array as follows:

```
MetalData(2).Type='Copper';
MetalData(2).SpecGrav=8.96;
MetalData(2).ThermCond=401;
MetalData(2).Resistivity=16.78E-9;
```

Now `MetalData` is a 1×2 structure array containing information about the two metals, tin and copper.

We can create multiple rows in a structure array by using double index notation, such as `MetalData(2,1)`.

We can add fields to the two structures within the MetalData in a similar manner.

```
MetalData(1).Symbol='Zn';
MetalData(1).Isotopes=[66 67 68];
MetalData(2).Symbol='Cu';
MetalData(2).Isotopes=[63 65];
```

Note that we had to use an index value of 1 to add fields to the first structure (Zinc), since there are now two structures in `MetalData`. When we first defined this single structure (for Zinc) there was no ambiguity, so an index was not needed, although it could have been used.

NOTE: We have used the field `Isotopes` to list only the stable isotopes. There are other isotopes of these metals with varying half-lives.

If we type `MetalData` in the command window, it will print the following to the screen:

```
MetalData =
1x2 struct array with fields:
    Type
    SpecGrav
    ThermCond
    Resistivity
    Symbol
    Isotopes
```

This tells us how many structures there are in `MetalData` and how they are organized (one row and two columns) as well as the names of the fields within the structures.

If we wish to know the contents of one of the structures within `MetalData`, we can type `MetalData(N)`, where `N` is the number of the structure for which we wish to see the contents. For example, typing `MetalData(2)` prints the following to the screen:

```
       Type:  Copper'
   SpecGrav: 8.9600
  ThermCond: 401
Resistivity: 1.6780e-08
     Symbol: 'Cu'
   Isotopes: [63   65]
```

COMPREHENSION CHECK 16-15

Create a structure array named `Resistors` containing data on three resistors, including fields for Value, Power, Composition, and Tolerance. The three resistors should have the following specifications (enter numeric values only for Value, Power, and Tolerance, not the units):

- 100,000 ohms, ¼ watt, Metal Film, 0.1%
- 2,200,000 ohms, ½ watt, Carbon, 5%
- 15 ohms, 50 watts, Wire Wound, 10%

Extracting Data from Structure Arrays

Accessing the contents of a structure array is very straightforward. We do not have to worry about "contents" as we did with cell arrays since each field has its own data type.

Using the `MetalData` example above, if we wanted the type of the first metal, we would type

```
MetalData(1).Type
```

This returns the text string `Zinc`.

If we wanted the first two characters of the type of metal number 1, we would type

```
MetalData(1).Type(1:2)
```

This returns the text string `Zi`.

We can use calculations and functions directly with values extracted from structure arrays.

```
min(MetalData.Resistivity)
```

finds the minimum resistivity of all of the metals `MetalData`.

```
max(MetalData(1).Isotopes)
```

This returns 68, the mass number (number of protons and neutrons) of the heaviest stable isotope of the first metal, Zinc.

COMPREHENSION CHECK 16-16

Use the data stored in `MetalData` above to answer the following questions:

(a) Write a single line of code that will determine the mass in kilograms of one cubic meter of zinc and place the result in `MZn`.

(b) Assume the number of one of the structures in `MetalData` is stored in a variable `MNum`. Write a single line of code that will place the mass number of the lightest isotope of the corresponding metal in the variable `LtIso`.

Functions Used with Structure Arrays

Table 16-9 lists several functions specifically designed to use with structure arrays.

Table 16-9 Selected Functions Used with Structure Arrays

Function	Description
`isstruct`	Determines if a variable is a structure array
`isfield`	Determines if a name is a field within a specific structure
`struct`	Another way to create structures
`orderfields`	Sorts field names in specified order
`setfield`	Another way to specify the content of fields
`rmfield`	Remove a field from a structure array
`cell2struct`	Convert cell array to structure array
`struct2cell`	Convert structure array to cell array

EXAMPLE 16-34

`isfield` returns a logical variable. Logical variables will be covered in a later chapter.

What is stored in the variable by following commands?

Assume `MetalData` has already been defined as above.

`IsF1=isfield (MetalData,'Density');`
`IsF1` *contains 0 (false), since density is not a field defined in* `MetalData`

`IsF2=isfield (MetalData,'SpecGrav');`
`IsF2` *contains 1 (true), since specific gravity is not a field defined in* `MetalData`

To create a structure array named `MetalSort` with the same data but with the field names sorted in alphabetical order,

```
MetalSort=orderfields(MetalData);
```

If you now type **MetalSort** in the command window, the following will appear:

```
1×2 struct array with fields:
        Isotopes
        Resistivity
        SpecGrav
        Symbol
        ThermCond
        Type
```

Note that this is an ASCII sort. In ASCII, capital letters have a "lower" value than lowercase values. If the field containing the element symbol had been lowercase – "symbol" – it would have appeared last in the list.

Note that the original structure array `MetalData` remains unchanged. A sorted copy was created and placed in `MetalSort`. Fields do not have to be sorted in ASCII order; other types of sorts, including custom sorts, are possible.

16.8 SAVING AND RESTORING VARIABLES

For a variety of reasons, we sometimes would like to be able to save all or some of the variables in the workspace and then restore them later. For example, we might wish to archive results, we might want to use the same data in another program, or we might need to share a set of variables with colleagues. Some of the online files associated with this text were created using the following method.

Variables can also be saved as .txt files in ASCII format, but this is beyond the scope of this textbook. See the MATLAB documentation files for more information.

WARNING

If your workspace already has variables with the same names as any of the variables being loaded, they will be replaced, thus losing the contents in those variables prior to the load. No warning will be generated.

To save all variables in the workspace, type

save ('filename')

In this case, the filename must have an extension of .mat on the end. If you do not include .mat, MATLAB will add it for you.

As an example, the command

```
save('TestVar.mat')
```

will save all of the variables currently in the workspace in a file named `TestVar.mat` placed in the current directory.

To place a set of stored variables into the workspace, type

load ('filename')

Again, the filename must have an extension of .mat on the end. For example,

```
load('TestVar.mat')
```

will place all of the variables stored in `TestVar.mat` into the current directory.

To save only selected variables in a .mat file, use the format

save ('filename', 'Var1', 'Var2', . . .)

As an example:

```
save('TV2.mat','Dens','Temp','Vol')
```

will store the three workspace variables `Dens`, `Temp`, and `Vol` in the file `TV2.mat`.

You can also load only selected variables from a .mat file. The format is similar to the save format. For example, assume `TV2.mat` was created as in the previous example.

```
load('TV2.mat','Dens','Vol')
```

will place the two variables `Dens` and `Vol` from the file `TV2.mat` into the workspace.

The commands `save` and `load` also support "wild card" variable names using the asterisk (*) symbol.

For example,

```
save('WC1.mat','Q*')
```

will save all workspace variables beginning with Q in the file `WC1.mat`.

```
load('WC2.mat','*3')
```

will load all variables stored in the file `WC2.mat` that end with 3.

```
save('WC3.mat','*_*')
```

will save all workspace variables containing an underscore character in the file `WC3.mat`.

COMPREHENSION CHECK 16-17

(a) Store all workspace variables in the file `TempData.mat`.
(b) Load all variables stored in `PressData` into the workspace.
(c) Save the workspace variables `Pr1`, `Pr2`, and `Tmp3` in the file `PTData.mat`.
(d) Store all variables whose names end with "`a1`" in the file `a1Var.mat`.
(e) Load all variables stored in `VelData` whose names contain "`XS`" into the workspace.

In-Class Activities

ICA 16-1

Which of the following are not valid MATLAB variable names? Circle all that apply. For those that are invalid, state why.

(a) `BigNum.docx`
(b) `Two_Nums`
(c) `This is a very long MATLAB variable name`
(d) `2BR02B`
(e) `Is_This_Valid?`
(f) `Var = V + 1;`
(g) `Mult2`
(h) `exp`
(i) `GClefSign`
(j) `AbcDEFGHijKLmnopqrstuvWXyz_Has_Lots_of_Characters`
(k) `DoNotPassGo_DoNotCollect$200`

ICA 16-2

Do the following scripted exercise. The >> prompt indicates what you should type into the MATLAB Command Window. Note that some of these commands may result in error messages.

`>> m=[10,20]*30`	% Multiply a row vector containing 10 and 20 by 30, store in variable m
`>> m1=[10;20]*30`	% Multiply a column vector containing 10 and 20 by 30, store in variable m1
`>> n=m+1`	% Add 1 to every element in variable m, store in variable n
`>> r=m+n`	% Add row vector in variable m to column vector in variable n, store in r. Note that even though this is dimensionally inconsistent, this works in MATLAB because it will allow any vectors (row or column) to be added together as long as they contain the same number of elements
`>> v=[2:2:6]`	% Create a row vector of numbers between 2 and 6, increasing by 2, store in v
`>> u=[1:0.5:3]`	% Create a row vector of numbers between 1 and 3, increasing by 0.5, store in u
`>> j=u+v`	% Add vector v to vector u Why will this not work?
`>> a=sin(u)`	% Take the sine of each element in the vector u
`>> b=sqrt(u)`	% Take the square root of each element in the vector u
`>> c=u^2`	% Square vector u Why will this not work?
`>> d=u.^2`	% Square each element of vector u
`>> e=[10,20;30,40]`	% Create a matrix of numbers 10 and 20 in the first row, 30 and 40 in the second row, store in the variable e
`>> f=[1,2,3;4,5]`	% Create a matrix of numbers 1, 2, and 3 in the first row, 4 and 5 in the second row, store in the variable f. Why will this not work?

ICA 16-3

Writing the MATLAB code necessary to create the following variables (parts (a)–(d)) or calculate the following matrix computations (parts (e)–(m)). If a calculation is not possible, explain why. You may assume that the variables created in parts (a) through (d) are available for the remaining computations in parts (e) through (m). For parts (e)–(m) that is possible, determine the expected result of each computation by hand.

(a) Save matrix $\begin{bmatrix} 3 & 8 \\ 5 & 7 \end{bmatrix}$ in variable W

(b) Save matrix $\begin{bmatrix} 4 & 8 & 1 \\ 5 & 1 & 6 \end{bmatrix}$ in variable X

(c) Save matrix $\begin{bmatrix} 3 \\ 5 \end{bmatrix}$ in variable Y

(d) Save matrix $\begin{bmatrix} 4 & 2 \\ 2 & 9 \end{bmatrix}$ in variable Z

(e) Add W + Z, save in variable A

(f) Subtract Y − Z, save in variable B

(g) Multiply W * X, save in variable C

(h) Multiply X * Z save in variable D

(i) Transpose X, save in variable E

(j) Multiply W * Z term-by-term, save in variable F

(k) Square Z, save in variable G

(l) Square X, save in variable H

(m) Add 20 to each element in X, save in variable I

ICA 16-4

Assuming `t=[9 10;11 12]` and `v=[2 4;6 8;10 12]` are currently stored in MATLAB's workspace, what is the output of each of the following statements? If an error will occur, explain why.

(a) `A = t(1,1) + v(1,1)`

(b) `B = t[1,1] + v[1,1]`

(c) `C = t * v`

(d) `D = v * t`

(e) `E = v(:,2) + t(:,2)`

(f) `F = v(2,:)+ t(2,:)`

(g) `G = t(1,1) * v(3,:)`

(h) `H = t(1,:) * v(1,:)`

(i) `J = t(1,:).* v(1,:)`

ICA 16-5

For each of the following problems, write a single MATLAB statement that will accomplish the stated purpose.

(a) Assume four scalars `S1, S2, S3,` and `S4` have already been defined. Place the smallest of these four values into `SmallS`.

(b) Assume that a vector `V1` has already been defined. Place the largest value in `V1` into `BigVal`.

(c) Assume three vectors `Va, Vb,` and `Vc` have already been defined. Find the length of the vector with the most elements and place the result in `MaxL`.

(d) Create a 500-element column vector `P17M11` containing two hundred number seventeens followed by three hundred number negative elevens.

(e) Create a vector `UV` that contains one of each value that is stored in another vector `V`, eliminating duplicates. Example: `V = [1 3 5 3 4 1]; UV = [1 3 4 5]`

(f) Create a vector `r100` that contains 100 random values between −5 and 5.

ICA 16-6

For each of the following problems, write a single MATLAB statement that will accomplish the stated purpose.

(a) Create a 150 × 150 matrix `N9999` that contains all zeros except for the major diagonal (upper left to lower right). All elements on the major diagonal should contain the value −9999.

(b) Create a 1200 × 1200 matrix `p75` that contains all zeros except for the minor diagonal (lower left to upper right). All elements on the minor diagonal should contain the value 75.

(c) Place the number of rows of matrix Ma into `MaRows` and the number of columns of matrix `Ma` into `MaCols`.

(d) Assume matrix `Mx` contains a single non-zero value. Place the row number containing this non-zero value into `NZR` and the column value in `NZC`.

(e) Assume matrix `Mz` has already been defined. Place the row numbers of all elements in `Mz` that are negative into vector `MzRN` and the corresponding column numbers into `MzCN`.

(f) Assume that matrix `MS` has already been defined. Place the elements on the major diagonal into row vector `MSD`.

ICA 16-7

For each of the following problems, write a single MATLAB statement that will accomplish the stated purpose. Assume a text string `TS1` has already been defined.

(a) Create a vector `LetLoc` with the same number of elements as `TS1`. For each element of `TS1` that is alphabetic, the corresponding element of `LetLoc` should equal 1. All other elements of `LetLoc` should be 0. Example: `TS1='%@3Gb6'` returns `LetLoc=[0 0 0 1 1 0]`

(b) Create a vector `Alphas` that has the numeric positions of all alphabetic characters in `TS1`. Example: `TS1='%@3Gb6'` returns `Alphas=[4 5]`

(c) Place the numeric index of the first alphabetic character in `TS1` into `Alpha1`. Example: `TS1='%@3Gb6'` returns `Alpha1=4`

(d) Place the last alphabetic character in `TS1` into `LastLetter`. Example: `TS1='%@3Gb6'` `LastLetter=b`

(e) Find all occurrences of the string `@3G` in string `TS1` and place the locations of the first character of each such substring (`@`) into `Str3G`. Example: `TS1='%@3Gb6kl@3G9@33G'` returns `Str3G=[2 9]`

ICA 16-8

Assume a structure array `SA1` has already been defined.

For each of the following problems, write a single MATLAB statement that will accomplish the stated purpose. You may use variables created in any problem to help solve problems farther down the list.

(a) Place the number of characters in the text string into `LenTex`.
(b) Place the number of alphabetic characters in the text string into `NumChar`.
(c) Place the number of non-alphabetic characters in the text string into `NumNon`.
(d) Place the number of values in the vector into `LenVec`.
(e) Place the sum of all elements in the vector except the last two into `SumM2`.
(f) Place the number of rows of the matrix into `MR` and the number of columns of the matrix into `MC`.
(g) Create a new 2 × 2 cell array `CA1SubM` from the matrix that contains the same contents as the matrix in `CA1` but divided into four sections:
- A scalar equal to the top left element from `CA1`
- A row vector equal to the remainder of the first row
- A column vector equal to the remainder of the first column
- A matrix with all remaining elements.

ICA 16-9

You wish to design a structure array to maintain a list of small hardware needed for a variety of different products. The structure should include the following information:

- An alphanumeric product code for each product being produced
- For each product, the general types of hardware needed, such as bolt, screw, washer, nut, etc.
- For each specific hardware type, the specifications, including (as appropriate) length, diameter, material, subtype, thread type, etc. Note that there may be more than one specification for a given type of hardware, such as bolts with two different lengths and diameters.
- For each specific hardware item, the quantity required for the product in question.

Create a structure array containing the parts lists for the two products below.

Product 1: Product code WP42

- 4 – steel hex-head bolts, 3/4 inch long, 1/8 inch diameter
- 8 – steel nuts 1/8 inch diameter

Product 2: Product code ES65

- 16 – steel torx-head bolts, 1 inch long, 5/16 inch diameter
- 8 – steel lock washers 5/16 inch diameter
- 16 – steel nuts 5/16 inch diameter

Chapter 16 REVIEW QUESTIONS

1. Assume four row vectors named `Prod10`, `Prod11`, `Prod12`, and `Prod13` contain data on production of various electronic devices at your company during the four years 2010, 2011, 2012, and 2013, respectively. The corresponding elements in each represent the number of a specific part manufactured during that year. For example, the first element might contain the number of 2N3904 transistors produced during each year, whereas the fifth column might contain the number of IC555 timer chips produced. You may assume all four vectors contain the same number of elements, corresponding to the same produced items.

   Write a single line of code to answer each of the following questions. You may use the results of any question to answer subsequent questions if desired. Note that your solutions should work regardless of the number of elements in the four vectors.

   (a) Create a new vector `TotalProd` that contains the total number of each item produced during the four-year period. Note that `TotalProd` will have the same number of elements as the original four vectors.

   (b) Create a new vector `AvgProd` that contains the average number produced per year of each item during the four-year period. Note that `AvgProd` will have the same number of elements as the original four vectors.

   (c) Create a four element column vector `YearProd` that contains the total number of all units produced during each year. 2010 production should be in the first (top) element.

   (d) Determine the maximum number of any type of device produced during each year and place the results in a four element column vector `MaxProd`.

   (e) Determine the maximum number of any device produced during any year and place the result in the scalar `OverallMax`.

   (f) If your company makes a profit of one-fifth of one cent on each device produced, regardless of type, determine the total profit made during this four year period and place the result in `Profit`. Your result should be in dollars.

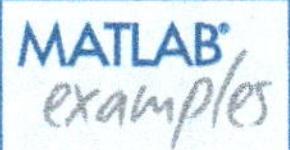

2. Assume a matrix named `Prod` contains data on production of various electronic devices at your company during several years. Each row of the matrix contains production data for a single year. The first element in each row contains the year, e.g., 2007 or 10012. The remaining elements in each represent the number of a specific part manufactured during that year. For example, the second element might contain the number of 2N3904 transistors produced during each year, whereas the fifth column might contain the number of IC555 timer chips produced. You may assume that corresponding elements in each row contain production numbers for the same type of device.

   Write a single line of code to answer each of the following questions. You may use the results of any question to answer subsequent questions if desired.

   A sample `Prod` matrix is provided online. Note that your solution must work for any properly formatted matrix `Prod`.

   (a) Create a row vector `TotalProd` that contains the total number of years in the first element and the total number of each item produced during all listed years in the remaining elements. Note that `TotalProd` will have the same number of elements as the number of columns in the `Prod` matrix.

   (b) Create a row vector `AvgProd` that contains the total number of years in the first element and the average number of each item produced during all listed years in the remaining elements. Note that `TotalProd` will have the same number of elements as the number of columns in the `Prod` matrix.

   (c) Create a two-column matrix `YearProd`. The first column should contain the same years as those in the first column of `Prod`, and the second column should contain the total number of all units produced during each year.

(d) Create a two-column matrix `MaxProd`. Determine the maximum number of any type of device produced during each year and place the results in the second column of the corresponding row in `MaxProd`.

(e) Determine the maximum number of any device produced during any year and place the result in the scalar `OverallMax`.

(f) If your company makes a profit of one-fifth of one cent on each device produced, regardless of type, determine the total profit made during all listed years and place the result in `Profit`. Your result should be in dollars.

(g) The solution to this problem is considerably more complicated than the corresponding problem using vectors (Review Question 1). What is the major advantage gained by this extra complexity?

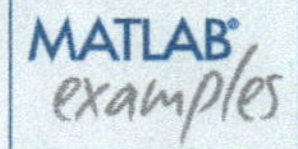

3. Assume that a cell array `PrCA` already exists, and contains a single item: the matrix `Prod` from Review Question 2. None of the other results required by that problem have been determined, however.

Augment `PrCA` so that it contains the original matrix `Prod` as well as all of the results required by Review Question 2. You may choose any organization of the cell array that seems logical to you. The information stored in each variable in Review Question 2 should be added to the cell array with a single line of code. You may NOT simply use the `struct2cell` function with the result from Review Question 4.

A sample `PrCA` cell matrix is provided online. Note that your solution must work for any properly formatted initial matrix in `PrCA`.

The sample matrix provided has data for five different devices (columns 2 through 6). Add a five element cell array to `PrCA` that contains the following part designations, corresponding to the five devices listed: `2N3904`, `2N3906`, `2N2222`, `IC555`, `IC741`. This should be the first element (first row, first column) of `PrCA`, thus the other entries will be shifted to other locations. Note that this is specific to the sample matrix provided. Other matrices would, of course, have different part designations.

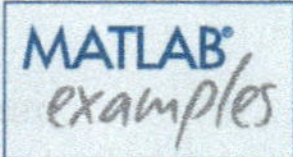

4. Assume that the matrix `Prod` from Review Question 2 already exists, but none of the other results required by that problem have been determined.

Create a single structure array named `PrStr` that contains the original matrix `Prod` as well as all of the results required by that problem. You may choose any organization of the structure array that seems logical to you as well as any field names that seem reasonable to you. You may NOT simply use the `cell2struct` function with the result from Review Question 3.

The sample matrix provided has data for five different devices (columns 2 through 6). Add a five element cell array to `PrStr` that contains the following part designations, corresponding to the five devices listed: `2N3904`, `2N3906`, `2N2222`, `IC555`, `IC741`. You may choose any field name you deem appropriate. Note that this is specific to the sample matrix provided. Other matrices would, of course, have different part designations.

5. (a) Create a cell array named `Cylinder` containing one row with the following elements:

- The text string Height(cm) The value 12
- The text string Diameter(cm) The value 1.5
- The text string Specific Gravity The value 3.5

(b) Create a cell array named `Pipe` containing the following elements:

- The text string Height(cm) The value 20
- The text string OD/ID(cm) A row vector containing the values 2.5 and 2.2

(c) Create a 2×3 cell array named `Parts` containing the following elements:

- The first row should contain the text string "Quantity", the value 100 and the cell array `Cylinder`
- The second row should contain the text string "Quantity", the value 15 and the cell array `Pipe`

(d) Using the `Parts` cell array, create the following variables. Note that your code should work if the contents of `Parts` changes—DO NOT hard code any values.

- `NumCyl`, containing the number of cylinders (1st row, 2nd column of `Parts`)
- `VolCyl`, containing the volume (cm^3) of one cylinder ($V = \pi r^2 H$), where the diameter is in the 4th element of the cell array in the 1st row, 3rd column of `Parts` and Height is in the 2nd element of that same cell array in `Parts`.
- `MassCyl`, containing the mass in grams of one cylinder. (You should be able to figure out where the specific gravity is by now.)
- `NumPipe`, containing the number of pipes (2nd row, 2nd column of `Parts`)
- `IVolPipe`, containing the internal volume (cm^3) of one cylinder ($V = \pi r^2 H$). You will have to determine where the values needed are stored using the explanations for `NumCyl` and `VolCyl` as a guide. Note that OD/ID means Outside Diameter/ Inside Diameter.
- Add two columns to the cell array `Parts`. The 4th column of `Parts` should contain the text Total Mass(g) in the 1st row and Total Volume(cm^3) in the 2nd row. The 5th column of `Parts` should contain the corresponding TOTAL mass of all of the cylinders and the TOTAL internal volume of all of the pipes.

6. Refer to the specifications for Review Question 16-5.

Create a structure array to contain the same information given in parts (a) through (c) in Review Question 16-5. Note that it may be appropriate to incorporate the text values into the structure array as field names instead of data. You may use any organizational structure and field names that seem logical to you—there are various possibilities.

Add entries to the structure array corresponding to the calculated volumes and masses listed in part (d) of Review Question 16-5. Again, you may use any organizational structure or field names that seem logical to you.

CHAPTER 17
PROGRAMS AND FUNCTIONS

Programming is the process of expressing an algorithm in a language that a computer can interpret. To correctly automate a process on a computer, a programmer must be able to correctly speak the language both grammatically and semantically. In this text, we communicate with a computer by using the MATLAB programming language. In the previous sections, the idea of creating algorithms to represent a process and creating variables in a programming language like MATLAB has been established, so this section intends to combine the two ideas to create programs and functions.

17.1 PROGRAMS

LEARN TO:
Understand how to navigate the MATLAB interface
Remember the naming rules for programs, functions, and variables
Create, execute, and modify programs in MATLAB

Certain principles apply to programming, regardless of what language is being used. Entire textbooks have been written on this subject, but a few important concepts are addressed here.

Notes About Programming in General

- **Programming style:** Just like technical reports should follow a particular format to ensure they can be understood by someone reading them, all programs should have a few common elements of programming style, particularly the inclusion of comments that help identify the source of the program and what it does. For any program, a proper header should be written describing the scope of the problem, including a problem statement and definition of all input and output variables used in the program. Properly commenting the source code is also critical for ensuring that someone else can follow your work on the program.
  As a rule of thumb, for each rectangle and diamond in a flowchart, there should be a few comments in the source code where that action occurs.
- **Program testing:** When you write a program, testing it is a critical step in making sure that it does what you expect. Lots of things can go wrong.
  - *Syntax errors*: These violate the spelling and grammar rules of the programming language. Compilers (programs that interpret your computer program) generally identify their location and nature.

- *Runtime errors*: These occur when an inappropriate expression is evaluated during program execution. Such errors may occur only under certain circumstances. Examples of runtime errors are division by zero, the logarithm of zero, or the wrong kind of input, such as a user's entering letters when numbers are expected. Computer programs should anticipate runtime errors and alert users to the error (rather than having the program unexpectedly terminate).
- *Formula coding errors*: It is not uncommon to enter a formula incorrectly in a form that does not create either a syntax or a runtime error. Evaluating even a few simple calculations by hand and comparing them to the values calculated by the program will usually reveal this kind of error.
- *Formula derivation errors*: This difficult error to discover in programming occurs when the computer is programmed correctly but the programming is based on a faulty conceptualization. This kind of error is difficult to figure out because programmers generally assume that a program that executes without generating any errors is correctly written. This situation underscores the importance of checking a solution strategy before starting to write a computer program.

- **Keeping track of units:** Some computer programs can be designed to have the user keep track of units as long as they use consistent units. Other computer programs require input data to be in a particular set of units. Regardless of which approach you use, the way in which units are managed must be clear to the user of the program and in the program's comments.
- **Documentation:** In any programming language, there is usually a special character or pair of characters that tell the program interpreter to ignore everything to the right of the special character to allow the programmer to write human readable notes within the code of the program. These notes are commonly referred to as "comments" within the code. A great deal of controversy surrounds how much documentation is necessary in any program, but this text proposes and follows the following guideline to properly document a program:
  For every numbered item in a written algorithm, you should include that item as a comment in your code.

In addition, at the top of each program, it is extremely helpful to at least write a problem statement and document any variables used within the program. It is helpful to cluster your description of variables into three groups: input variables, output variables, and function variables.

Notes Particular to MATLAB

Before proceeding with the material on MATLAB, it would be wise to review the material on matrices in the appendix materials. Since variables in MATLAB are inherently matrices, it is critical that you understand not only some basic matrix concepts, but also the specific notation used in this text (as well as MATLAB) when referring to matrices. Certain details specific to MATLAB are important to note early on.

- **The MATLAB interface:** The MATLAB program window, shown below, has a number of subwindows. The two main windows we are concerned with for now are the Command Window and the Editor Window.

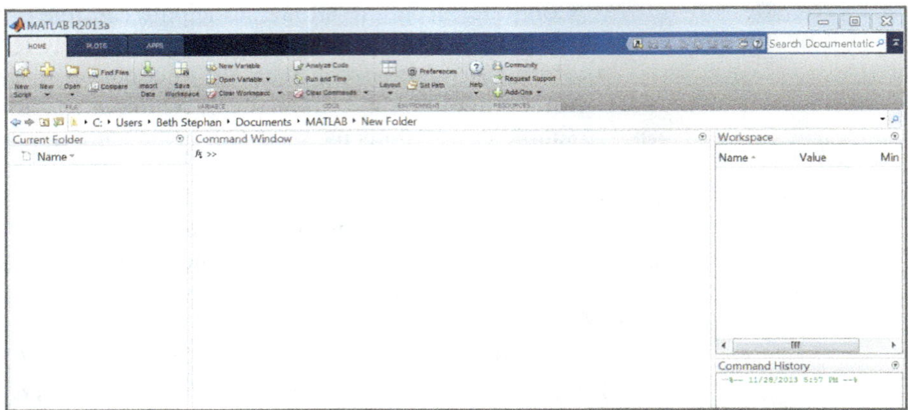

- The *Command Window* can be used as a powerful but awkward calculator. The button at the top-right side of the Command Window menu bar will display a menu for common layout commands, including the ability to close or "undock" the Command Window from the MATLAB program window.

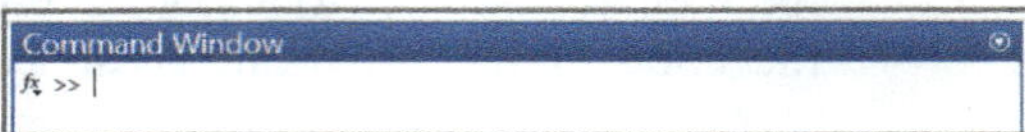

To restore the appearance of MATLAB to the default layout, select the **Layout** menu from the **Home** ribbon and click **Default**.

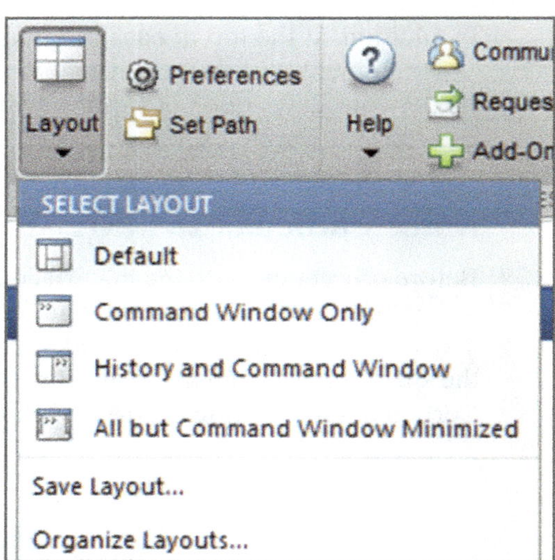

NOTE
`clc` = clear the Command Window

To clear any text from the Command Window, enter the special reserved MATLAB expression `clc` in the Command Window.

FILE TYPES

Word = .docx

Excel = .xlsx

Powerpoint = .pptx

MATLAB = .m

- Although you can enter MATLAB statements in the Command Window, the *Editor/Debugger window* is the interface of choice for developing MATLAB programs. The editor stores your commands in M-files. This permanent record of your commands allows you to make small changes to either the program or the inputs and to re-execute a complicated set of commands. The editor also assists in the formatting of MATLAB programs, using both indentation and color to distinguish program elements. To launch the editor window, press the **New Script** button on the **Home** ribbon or use CTRL+N. The Editor interface will launch in a new window.

- **Naming programs, functions, and variables in MATLAB:** In MATLAB, program names, function names, and variable names must all follow certain rules to prevent unintended side effects:

Naming Rules

- All names must consist only of letters, numbers, and the underscore character (displayed by holding "Shift" and typing a hyphen).
- Names can *only* begin with alphabetic letters (no numbers or special characters).
- Names cannot be longer than 63 characters.
- Names in MATLAB are case sensitive: `Frog` and `frog` are different.
- Names cannot have the same name as any other identifier (program name, function name, built-in function, built-in constant, reserved word, etc.), because MATLAB's order of execution will prevent the program from running in the intended order.

COMPREHENSION CHECK 17-1

Which of the following are valid MATLAB variable names?

(a) `my name`	(d) `m`	(g) `my_var`	(j) `for`
(b) `length`	(e) `m6`	(h) `@clemson`	(k) `HELLO[]`
(c) `MyLength`	(f) `4m`	(i) `my.variable`	

NOTE

Order of Search:

- Variable
- Built-in function
- Program or function in current directory
- Program or function in current path

- **Order of Search:** Order of search is the sequence MATLAB goes through when you type a name in either the Command Window or in a program or function. If you type `Frog` in the Command Window:

1. MATLAB checks if `Frog` is a variable, and displays its value if it is.
2. MATLAB checks if `Frog` is one of MATLAB's built-in functions, and executes that function if it is.
3. MATLAB checks if `Frog` is a program or function in its current directory, and executes the program or function if it is.
4. MATLAB checks if `Frog` is a program anywhere in its path, and executes the program or function if it is. The location where you store your MATLAB programs or functions must be included in MATLAB's path. You can check where MATLAB is looking for programs and functions by typing `path` in the Command Window. Any M-files in the directories listed will be visible to MATLAB. If you

try to execute an M-file outside any of the directories listed, MATLAB will ask if you want to switch current directories or add that directory to the list of paths MATLAB can see.

- **what and who:** To display a list of all MATLAB programs in the current directory, type `what` into the Command window. To display all variables currently in MATLAB's workspace, type `who` in the Command Window. Remember that all variables automatically clear after you exit MATLAB.
- **Comments:** For documenting or commenting any code in MATLAB, a special character is used to "comment" out text: the percent symbol (%). The comment character can appear anywhere in a MATLAB program or function, but preferably either at the beginning of a line or at the middle of a line of MATLAB code. Everything after the percent sign to the end of the line is ignored by MATLAB, so you as a programmer can "comment out" any broken lines of code during code testing in addition to providing comments. Program or function file headers will usually have 6–10 lines of comments at the top of every program and more throughout the rest of the program or function.

```
% This is an example of a comment line.
% Comment lines are shown in the Editor Window in green.
```

A **block comment** is a block of non-executed text between the special block characters **%{** and **%}** that allow the user to type full length paragraphs as a comment without needing to type the % symbol before each sentence. For example:

```
%{
This is an example of using
a block of comments in a code
%}
```

Program Structure and Use

Programs are a set of instructions given to a computer to perform a specific task. In this section, we create MATLAB program files, or **M-Files**, to automate our processes. To create a program in MATLAB, launch the MATLAB editor. By default, the editor will create a new, blank document named Untitled, or Untitled2, or Untitled3, and so on, depending on how many unsaved programs are open in the editor at once. The following example provides a walkthrough for creating a basic program, as well as screenshots of the actions described in the text to help you to follow along while recreating these steps on your own computer running MATLAB.

EXAMPLE 17-1

We want to create a MATLAB program to convert Cartesian coordinates (x, y) to polar coordinates (r, θ). Before we can even attempt to write the program, we need to recall some information about the coordinate systems.

We know the following:

- In the Cartesian coordinate system, the coordinates (x, y) represent how "far away" a particular data point is from the origin $(0, 0)$ by discussing the horizontal and vertical distance with respect to the origin. The x-value represents the horizontal distance and the y-value represents the vertical distance.

- In the polar coordinate system, the coordinates (r, θ) represent the exact distance from the point to the origin (r) as well as the angle of elevation (θ) with respect to the origin (0, 0) in units of radians.
  - The conversion from polar coordinates to Cartesian coordinates is
    $$x = r\cos\theta$$
    $$y = r\sin\theta$$
    where θ the angle of elevation in radians.
  - The conversion from Cartesian coordinates to polar coordinates is
    $$r = \sqrt{x^2 + y^2} \qquad \theta = \tan^{-1}\left(\frac{y}{x}\right)$$

We have the equations necessary to write a program to convert Cartesian coordinates to polar coordinates, and vice versa.

First, we need to ask MATLAB for more information on doing things like calculating a square root and calculating an inverse tangent of a value.

In MATLAB's Command Window, we search for the built-in functions to calculate a square root and an inverse tangent. The `lookfor` *function allows you to search MATLAB help documentation for certain key words. Note that if you type* `lookfor square root` *into the Command Window, MATLAB displays an error message. The* `lookfor` *command is expecting only one keyword following the command "lookfor", so we decide to try our search again, this time including only the word "square."*

NOTE

`CTRL + C` = stops MATLAB from executing.

Because MATLAB is searching through every built-in function, we need a way to tell MATLAB to stop, because when the function `sqrt` *appears, it seems to be the exact function we need. To terminate a MATLAB command early, we press* **`CTRL + C`** *(i.e., the CTRL key and the "C" key at the same time) on our keyboard, and the prompt will return to the Command Window.* `CTRL + C` *can also be used to stop a program caught in an infinite loop or that otherwise wandered away and got lost.*

To ask for more information on the function `sqrt`*, we type* `help sqrt` *to double-check that the function is going to perform our intended calculation as well as details of how to use it.*

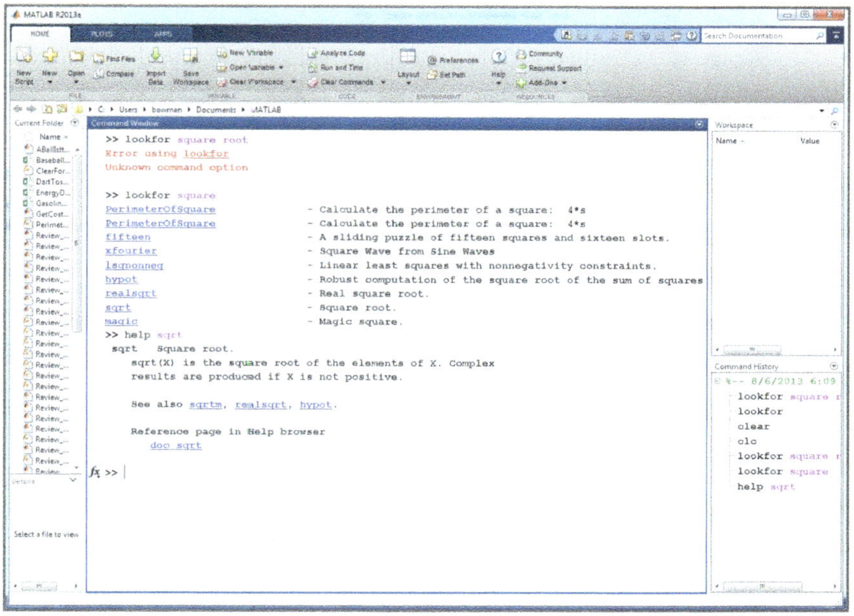

```
>> lookfor square root
Error using lookfor
Unknown command option

>> lookfor square
PerimeterOfSquare              - Calculate the perimeter of a square:  4*s
PerimeterOfSquare              - Calculate the perimeter of a square:  4*s
fifteen                        - A sliding puzzle of fifteen squares and sixteen slots.
xfourier                       - Square Wave from Sine Waves
lsqnonneg                      - Linear least squares with nonnegativity constraints.
hypot                          - Robust computation of the square root of the sum of squares
realsqrt                       - Real square root.
sqrt                           - Square root.
magic                          - Magic square.
>> help sqrt
 sqrt   Square root.
    sqrt(X) is the square root of the elements of X. Complex
    results are produced if X is not positive.

    See also sqrtm, realsqrt, hypot.

    Reference page in Help browser
       doc sqrt

fx >>
```

Repeating this process to find the inverse tangent function, we determine that `atan` *is the correct function necessary to calculate the inverse tangent. We note, from the help documentation, that the* `atan` *function will return the inverse tangent in radians.*

To create a program, we click the ***New Script*** *button on MATLAB's Home tab. After creating a new blank M-file, we need to write out our problem statement as code comments. Comments are any non-interpreted text that you would like to include in your code file. All comments start with the percent symbol (%) and everything to the right of the symbol will be ignored on a per-line basis. Comments can follow any executable statements in our code, but for now, we will add the first couple of lines in our MATLAB program by writing out a short problem statement with the % symbol before any text we write.*

Next, we start writing our code. In a MATLAB program, it is good practice to start the program by clearing all of the variables from the workspace (`clear`*) and clearing the Command Window (*`clc`*). That way, we guarantee that if we run our program on a different computer or run it after restarting MATLAB, our program will create all the variables we need to solve the problem and not rely on the workspace having variables predefined before code execution.*

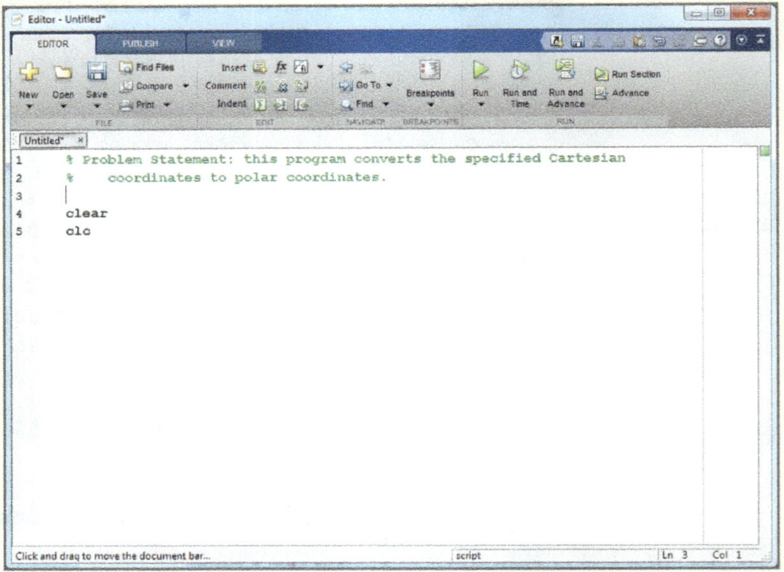

Now we are ready to start our program. Let us look at the step-by-step written algorithm necessary to solve this problem:

1. *Create a variable to contain the x-coordinate as a variable named "x."*
2. *Create a variable to contain the y-coordinate as a variable named "y."*
3. *Calculate the radius by taking the square root of the sum of the squares of x and y and store the result in a new variable called "r."*
4. *Calculate the angle by taking the inverse tangent of the result of dividing y by x and store the result in a new variable called "Theta."*

In the written algorithm, the first thing we have to do is to manually assign the values of x and y. For this example, we assign x to contain the value of 5 and y to contain the value of 10, but in later chapters, we show how the user of a program can enter values. Note that in addition to creating the variables x and y, we have also added additional comments in the header of our program documenting what these variables represent.

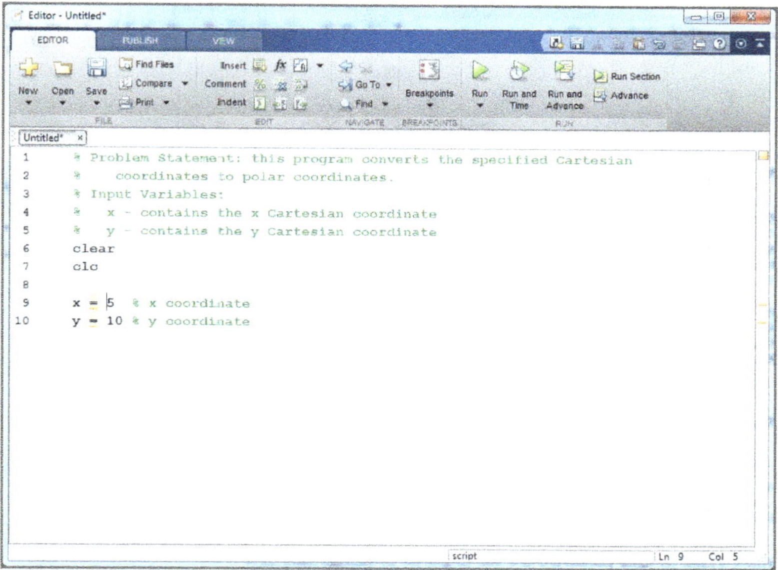

Note that MATLAB highlights the = symbol in the editor. When we move the cursor over the assignment operator (=), a pop-up message tells us that we can suppress the output of the assignment expression by using a semicolon. For now, we omit the semicolon and observe MATLAB's behavior when we leave the code unsuppressed.

Next, we type the equations to calculate the radius and the angle, using the built-in MATLAB functions we discovered with `lookup` *and* `help`*. In addition, we've added extra comments about the new variables representing the radius and angle of the polar coordinate.*

```
1  % Problem Statement: this program converts the specified Cartesian
2  %    coordinates to polar coordinates.
3  % Input Variables:
4  %   x - contains the x Cartesian coordinate
5  %   y - contains the y Cartesian coordinate
6  % Output Variables:
7  %   r - the radius component of the polar coordinate
8  %   Theta - the angle component of the polar coordinate
9  clear
10 clc
11
12 x = 5  % x coordinate
13 y = 10 % y coordinate
14
15 % calculate the radius polar coordinate
16 r = sqrt(x^2+y^2)
17
18 % calculate the angle polar coordinate
19 Theta = atan(y/x)
```

NOTE

Remember, MATLAB programs and functions *cannot* have any spaces in their file name, so be sure that a file is named without spaces, for example, CartesianToPolar.m—*not* "Cartesian To Polar.m."

Next, we save the program in MATLAB's path with a file name that follows the program-naming convention. We choose the name "CartesianToPolar" since that name nicely describes the purpose of the program.

To save a program, we click the ***Save*** *icon in the Editor window.*

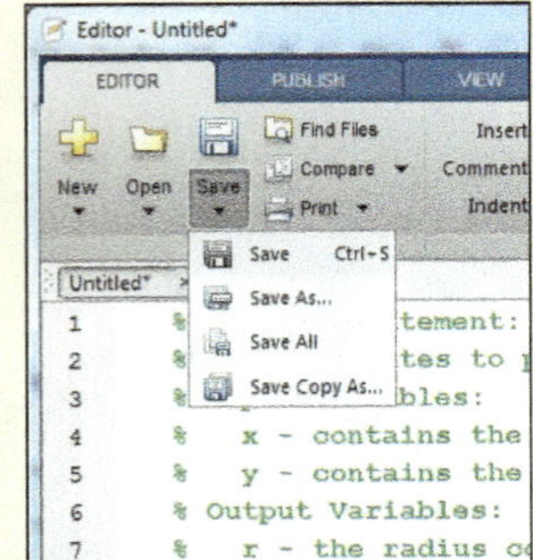

⌘ ***Mac OS:*** *MATLAB is one of the few programs that does not modify the main menu at the top of the screen. Instead, its main menu is at the top of the MATLAB window, just as in the Windows implementation. This is because MATLAB is actually running under an X Windowing System, similar to Linux or UNIX®.*

Running the Program

There are a number of different ways to run a MATLAB program located within MATLAB's path.

Method 1: *In the main MATLAB window (not the Editor), type the name of the program in the Command Window and press* ***Enter.***

Method 2: *In the main MATLAB window (not the Editor), right-click the program you want to run while in the Current Directory box and select* ***Run****.*

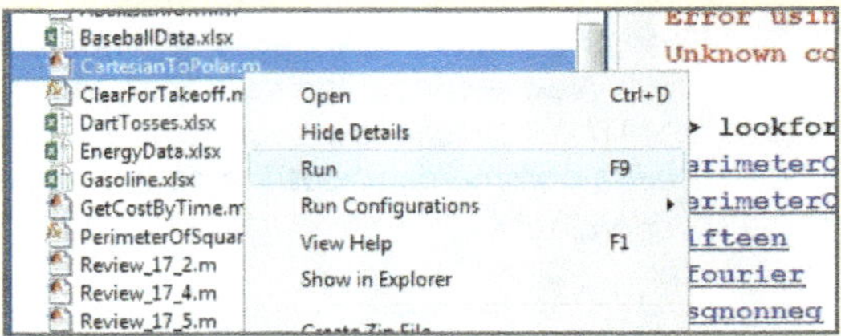

Method 3: *With the program loaded in the Editor window, click the* ***Save and Run*** *button in the toolbar.*

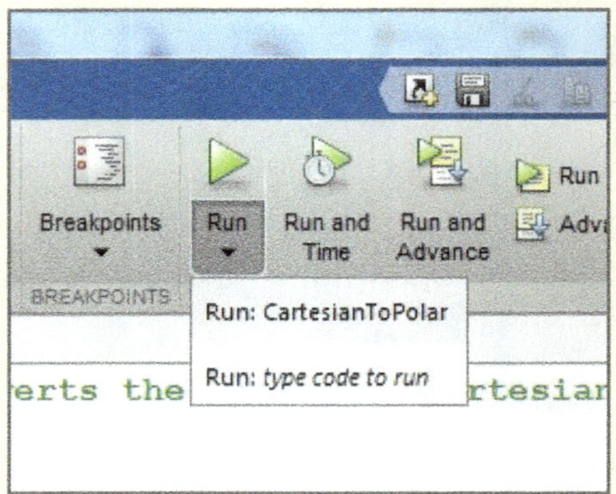

NOTE

To suppress output, end the code line with a semicolon (;).

We see that MATLAB spits out the results of every assignment and calculation performed in the program. Let us assume we want to show only the output from the calculations of r and Theta, and not the output of the assignment of x and y. To do this, we add a semicolon (;) at the end of each line of code we do not want showing up in the output of the program.

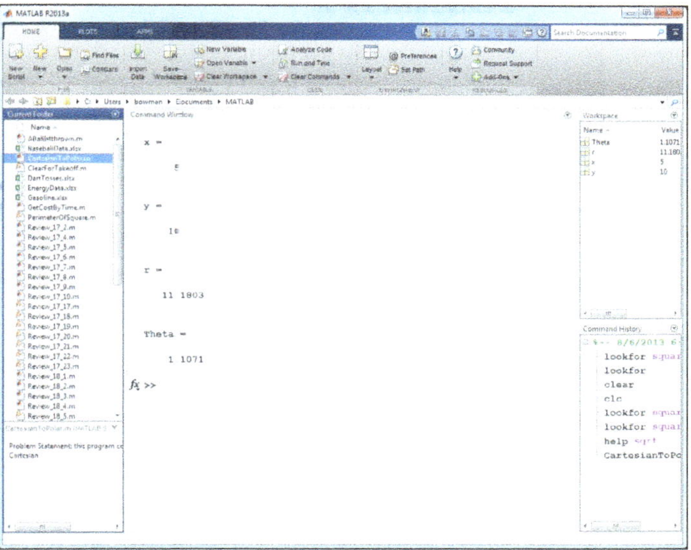

Now when we save and run our modified program, we will see only the output from the calculations of r and Theta.

Programming Features in Our Coverage of MATLAB

MATLAB has many advanced programming features that make sophisticated numerical analysis, including simulation, possible. In this text, you will learn basic programming structures to solve engineering analysis problems.

- **Data types:** Covered in the previous chapter, there are different data types that can be stored and manipulated by MATLAB. Some of the more common data types used by programs include numeric and string variables, but many applications also use vector and matrix variables for advanced calculations, as well as special types like cell arrays for storing a combination of different types within a single structure.
- **Input and output:** By gathering input, the program can solve a different problem each time the program is executed. Output statements allow you to produce formatted text to communicate with the user. Likewise, being able to read and write Microsoft Excel files using MATLAB code allows your program to perform calculations without needing to type in a bunch of data and permanently save results to special MATLAB data files (called .mat files.)
- **Conditional statements:** These allow a program to take different paths for various user input, intermediate calculations, or final results. Conditional statements allow the programmer to ask questions in order to make decisions.
- **Looping:** Repetition is critical in breaking down complicated problems into many simple calculations.
- **Plotting:** Graphs communicate data relationships in a variety of forms.

17.2 FUNCTIONS

LEARN TO: **Create, execute, and modify functions in MATLAB**
Understand the difference between local and global variables

Functional programming allows the designer of an algorithm to recycle a segment of code to reduce the amount of redundant code. Decreasing code redundancy allows for cleaner source code that is easier to read. **Functions** are a special class of programs that require the user to pass in input variables and capture output variables.

User-defined functions are similar to programs in that they contain a series of executable MATLAB statements in M-files that exist independently of any program we create in MATLAB. To allow any MATLAB program to call a user-defined function, we must follow a few guidelines to ensure that our function will run properly.

LAW OF ARGUMENTS

In computer programming, and often in math as well, the word "argument" does *not* refer to an altercation or vehement disagreement. In this context, it means the information that is given to a function as input to be processed. If you use the function sqrt to find the square root of 49, you would write `sqrt(49)`: 49 is the argument of sqrt—the value upon which the sqrt function performs its calculations. Functions may have zero arguments (e.g., `rand` will calculate a single random number between 0 and 1—no input is required), one argument (e.g., `sqrt(4)`), or two or more arguments (e.g., `power(a,b)` raises a to the power b).

Function Creation Guidelines

- The first line in a function must be a function definition line:

```
function [output_variables] = function_name(input_variables)
```

  - ***output_variables:*** A comma-separated list of output variables in square brackets.
  - ***input_variables:*** A comma-separated list of input variables in parentheses. The list of input variables to a function are commonly referred to as **arguments**.
  - ***function_name:*** Function names cannot be the same as a built-in function, user-created variables, or any reserved word in MATLAB. Avoid creating variables with the same name as a user-defined function because you will not be able to call your function.
- **The name of the function must be the same as the file name**. For example, if the function name is `sphereVolume`, the function must be saved in an M-file named `sphereVolume.m`, where `.m` is just the extension of the M-file. In addition to the aforementioned function-naming rules, the name of the function must also follow all program-naming rules, including omitting spaces in the file name.
- To execute a function, you must call it from MATLAB's Command Window or from a different program or function by passing in input variables and saving the resulting output into different variables.
- Do *not* use the built-in function `clc` within one of your own functions. The functions you are writing are intended to be general purpose and can be called from other

programs (or functions), so if text is written to the screen prior to calling a function that contains the `clc` command, it will wipe the Command Window clean—losing all of the previous output on the screen. For a different reason, the `clear` command should not be used within a function since functions usually contain input arguments, so calling the `clear` command would delete any input variables and their values. In general, the `clear` and `clc` commands should only appear at the top of programs, not functions!

Function Structure and Use

The following examples demonstrate the structure of different functions and how MATLAB programs can use the functions to reduce code redundancy.

EXAMPLE 17-2

Assume we are required to create a function, `areaCircle`, to calculate the area of a circle given the radius. The function should accept one input (radius) and return one output (area). Furthermore, we are told to suppress all calculations within the function.

Source Code

```
% Problem Statement: This function calculates the area of a
% circle
% Input:
%   r—Radius [any length unit]
% Output:
%   A—Area [units of input^2]
function [A]=areaCircle(r)
% Calculation of the area of a circle
A=pi*r^2;
```

Typical Usage (typed in Command Window or a separate program):

You can call the function by typing the name of the function and passing in a value for the radius. Note that the result is stored in a variable called "ans" when you do not specify a variable to capture the output.

```
>> areaCircle(3)
ans=
     28.2743
```

You can call the function by typing the name of the function and passing in a variable.

```
>> Rad=4;
>> CircleA=areaCircle(Rad)
CircleA=
   50.2655
```

COMPREHENSION CHECK 17-2

What is the output when you "pass in" the value 10 to the function in Example 17-2?

Assume that you want to store the result of the previous function call in a variable called `Dogs`. What is the command you would type in MATLAB?

COMPREHENSION CHECK 17-3

Write a function named RAC that will accept two parameters, N and A. The function should return a single value R calculated as `R = 3 sin(N^A)`. For this example, you do not need to include any comments with this code.

Testing Functions

MATLAB's Editor contains functionality to allow you to efficiently test functions that you write by allowing you to create different test scenarios and providing a shortcut to run those test cases you can create directly in the editor window. Assume we have created and saved the following function in MATLAB:

```
function [out1,out2,out3]=myFunctionInTheEditor(in1,in2)
out1 = in1 * 2;
out2 = out1 + in2;
out3 = [out1, out2]*3;
```

This function named `myFunctionInTheEditor` accepts two input arguments (`in1, in2`) and returns three output variables: `out1` and `out2` are both numbers and `out3` is a vector. Assume we want to test to see what the resulting output variables will contain when the value of 5 is used for `in1` and 10 is used for `in2`. To do this, we can use the same Run button we used for programs. The first time you press the Run button, an error message will appear in the screen indicating that the input variables are missing:

```
>> myFunctionInTheEditor
Error using myFunctionInTheEditor (line 2)
Not enough input arguments.
```

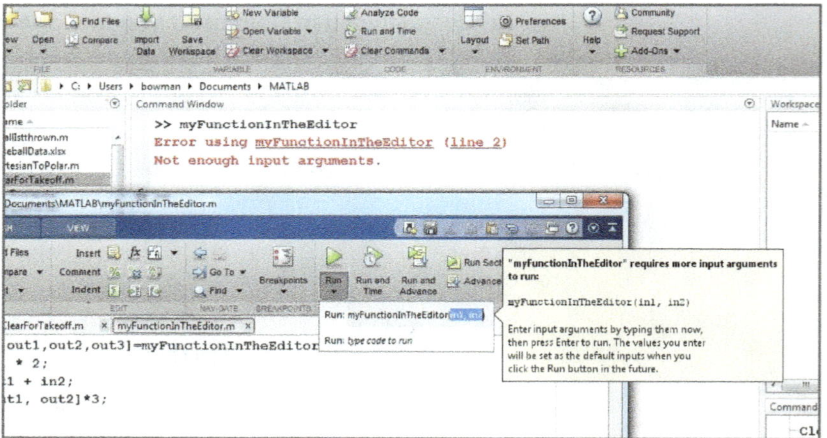

However, a little pop-up dialog appears in MATLAB's Editor Window that actually allows you type in values for the test case. Type in **5, 10** within the parenthesis on the dialog, representing the values of `in1` and `in2` respectively and press the Enter key. You'll notice that output has appeared in the Command Window:

```
>> myFunctionInTheEditor(5,10)
ans =
      10
```

This output shows that the function was successfully executed with the value of 5 for `in1` and 10 for `in2`, but since this function didn't save any output variables, the value 10 is stored in the default variable `ans`. By default, MATLAB requires you to provide values for all of the input variables, but it does not require you to capture all of the output variables with each call. The value stored in `ans`, 10, is actually the value of the `out1` variable calculated in the function. However, the Editor Window is smart enough to be to set up test cases where all of the output variables are assigned to variables when the function is called. Consider the test case with the value 3 for `in1` and 10 for `in2`. To save all three output variables (for testing, we will save `out1` into variable `a`, `out2` into variable `b`, and `out3` into variable `c`), select the Run drop down menu on the Editor Window again and this time type the following in the "type code to run" box: `[a,b,c] = myFunctionInTheEditor(3,10)`

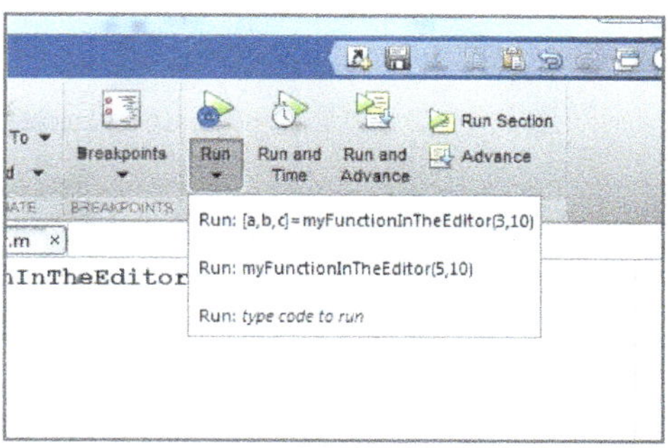

After executing the function with our new test case, the output variables are displayed properly in the Command Window:

```
>> [a,b,c]=myFunctionInTheEditor(3,10)
a =
     6
b =
     16
c =
     18  48
```

EXAMPLE 17-3

Assume we are required to create a function, `volumeCylinder`, to calculate the volume of a cylinder given the radius. The function should accept two inputs (radius, height) and return one output (volume). For this example, you may assume that the user will input the radius and height in the same units. Any calculations in the functions must be suppressed.

Algorithm

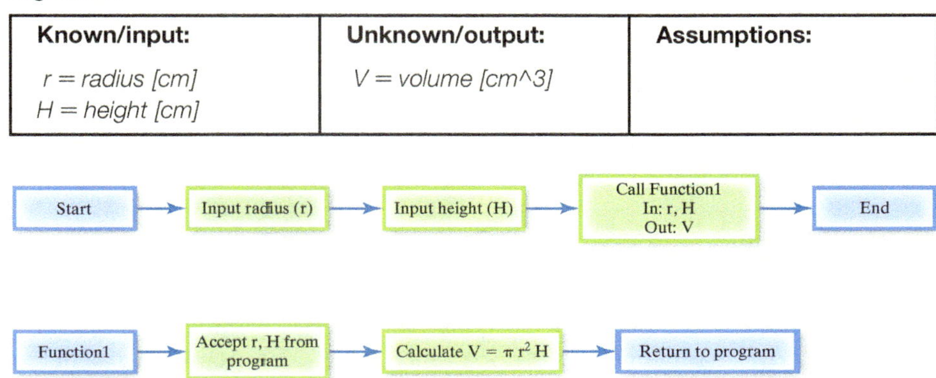

Known/input:	Unknown/output:	Assumptions:
r = radius [cm] *H = height [cm]*	*V = volume [cm^3]*	

Source Code

```
% Problem Statement: Calculate the volume of a cylinder.
% Input:
%   r-Radius [any length unit]
%   H-Height [same length unit as radius]
% Output:
%   V-Volume [units of input^3]
function [V]= volumeCylinder(r, H)
% Calculation of the volume of a cylinder
V=pi*H*r^2;
```

Typical Usage

```
>> Vol1=volumeCylinder(1,2)
Vol1=
   6.2832

>> Vol2=volumeCylinder(2,1)
Vol2=
   12.5664
```

Note that the order of arguments passed into the function matters. The order of variables expected by the function is based on the order they are listed in the function header. Therefore, the first input argument is the radius, followed by the height.

NOTE

Local variables are only visible to the program or function using them.

Global variables are visible within any program or function.

Local and Global Variables

One important fact to note about variables is that the variables in the workspace are only available within MATLAB's Command Window and to any running program. Functions create a private workspace and cannot access any of the variables available in MATLAB's main workspace. The only way to access any variables in MATLAB's main workspace is to "pass" the variable to the program as function input. The phenomenon of having "main workspace" variables inaccessible by functions is because functions exist within a different scope from the programs. The variables created within the "private workspace" of a function are commonly referred to as **local variables**, since they are destroyed after the function executes and do not appear in the "main workspace" unless they are passed back as function output. Special constants like "pi" exist within the scope of programs. Variables that are visible within any scope are commonly referred to as **global variables**. Global variables are created in functions using the reserved word **global** followed by a comma separated list of the variables that should be considered global variables.

```
function [x] = functionName(y)
global z
z=30;
```

In this example, after the function `functionName` is called, the value of `z` is accessible by other programs or functions.

In the diagram below, consider the two inner boxes to be programs (`Prgm1`) or functions (`Func1`) and the outer box represents all of the other variables in MATLAB's Workspace. Since we do not include any commands to clear the variables from the workspace at the top of the program `Prgm1`, the program is able to "see" and use all of the variables within MATLAB's workspace. In addition, it will create a new variable in MATLAB's workspace (c) after we execute program Prgm1.

In contrast, the function cannot "see" or use any of the variables in MATLAB's Workspace and creates its own separate workspace to execute the code inside the function. In order to run the function, the programmer must provide the values of variables e and g by "passing" the variables into the function, the code calculates the value of d, and returns d as function output. Assume that we store the output of the function in the variable h when the function returns to the MATLAB workspace. After the function has executed, the separate workspace is destroyed, along with variables d and k, which were never stored in MATLAB's main workspace.

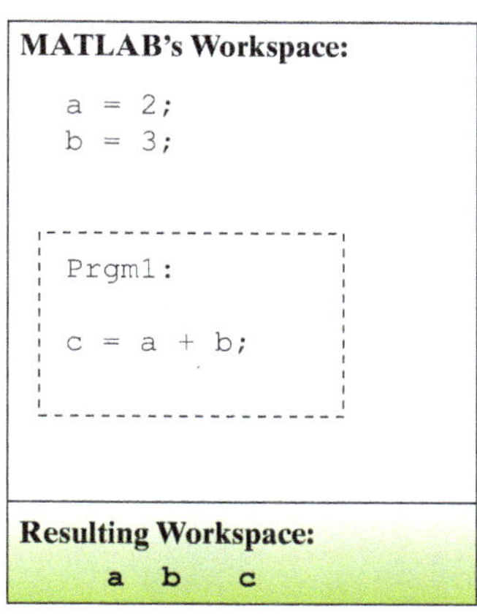

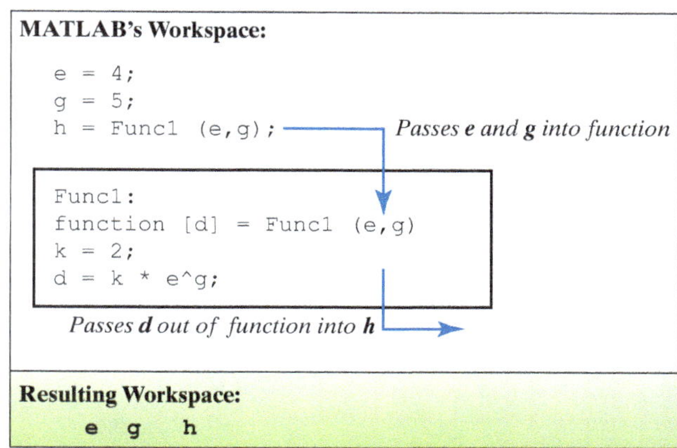

● EXAMPLE 17-4

Assume we are required to create a function, circleCalculations, to calculate the area and perimeter of a circle given the radius. The function should accept one input (radius) and return two outputs (area, perimeter). Any calculations in the functions must be suppressed.

Source Code

```
% Problem Statement: This function calculates the area and
% perimeter of a circle.
% Input:
%   r—Radius [any length unit]
% Output:
%   A—Area [units of input^2]
%   P—Perimeter [units of input]
function [A,P]=circleCalculations(r)
% Calculation of the area of a circle
A=pi*r^2;
% Calculation of the perimeter of a circle
P=2*pi*r;
```

Typical Usage

```
>> [Area,Perimeter]=circleCalculations(5)
Area=
   78.5398
Perimeter=
   31.4159
```

EXAMPLE 17-5

Assume we are required to create a function, `cylinderCalculations`, to calculate the volume and lateral surface area of a cylinder given the radius and height. The function should accept two inputs (radius, height) and return two outputs (volume, surface area). For this example, you may assume that the user will input the radius and height in the same units. Any calculations in the functions must be suppressed.

Source Code

```
% Problem: Function calculates volume and surface area of
% cylinder.
% Input:
%   r–Radius [any length unit]
%   H–Height [same length unit as radius]
% Output:
%   V–Volume [units of input^3]
%   SA–Surface area [units of input^2]
function [V,SA]=cylinderCalculations(r,H)
% Calculation of the volume of cylinder
V=pi*H*r^2;
% Calculation of the surface area of a cylinder
SA=2*pi*r*H;
```

Typical Usage

```
>> [Volume,Surface]=cylinderCalculations(3,2)
Volume=
  56.5487
Surface=
  37.6991
```

17.3 DEBUGGING MATLAB CODE

LEARN TO: Utilize the Debugger tool in MATLAB to eliminate errors
Examine the variable values during execution of a program or function

If a program has syntax errors, MATLAB (and other compilers) will give you useful feedback regarding the type and location of the error to help you fix, or **debug**, your program. If this information is not sufficient for you to understand the problem, check with others who might have seen this error previously—some errors are particularly common. If you still cannot identify the error, you might need to use a more formal process to study the error. For other kinds of errors, this formal process is essential for diagnosing the problems. When you first write a program, you must test the output of the program with a set of inputs for which you know result. Such test cases would include:

- Simple cases for which you can quickly compute the expected output;
- Cases provided by an instructor or textbook with a published solution;
- Cases for which results have already been produced by a previously tested program that does the same thing;
- Test cases that are customarily used to test programs.

NOTE

Why do we say that computers (and programs) have bugs? Grace Murray Hopper is quoted in the April 16, 1984, issue of *Time Magazine* as saying, "From then on, when anything went wrong with a computer, we said it had bugs in it"—referring to when a 2-inch-long moth was removed from an experimental computer at Harvard in 1947.

When the output of a program is different from what you expect, the debugging process begins. It is common to debug shorter programs simply by reading them and writing a few notes. Longer programs may require the use of the MATLAB Debugger, which is available as part of the MATLAB Editor. Whether you are using the MATLAB Debugger or are debugging "by hand" or in the Command Window, the same techniques apply.

Preparing for Debugging

When using the Editor/Debugger, open the program for debugging. If it is open, make sure changes are saved—MATLAB will run the saved version without including recent changes. If you are debugging from a printed program and output, make sure that the printed output came from the version you are reading. In preparing to debug a program, it is critical to be able to reproduce the conditions that caused the bug in the first place. If the bug occurs in processing data from a large data set, you may have to split the data set to find the specific data that triggered the bug.

Setting Breakpoints

NOTE

A breakpoint stops the program at the specified location to allow you to examine the content before the program continues.

When using the Editor/Debugger, establish **breakpoints** that allow you to check your agreement with the program at various stages. A breakpoint stops the program at the specified location to allow you to examine the contents of variables, etc., before the program continues. This will help you find the specific location when the program does something you were not expecting. When debugging shorter programs, write down everything the program does or make all program results display in the Command Window by removing semicolons that were used to suppress output to the Command Window.

MATLAB has two main types of breakpoints: standard (set by location in the program) and conditional (triggered at a specified location by specified conditions). We focus on standard breakpoints in this section. Standard breakpoints are normally shown as little red dots next to the line of code where the executing should pause and conditional breakpoints will appear as little yellow dots. If the breakpoint dot displays gray, either the file has not been saved since changes were made to it or there is a syntax error on the line or somewhere in the file. All breakpoints remain in a file until you clear (remove) them or until they are automatically cleared.

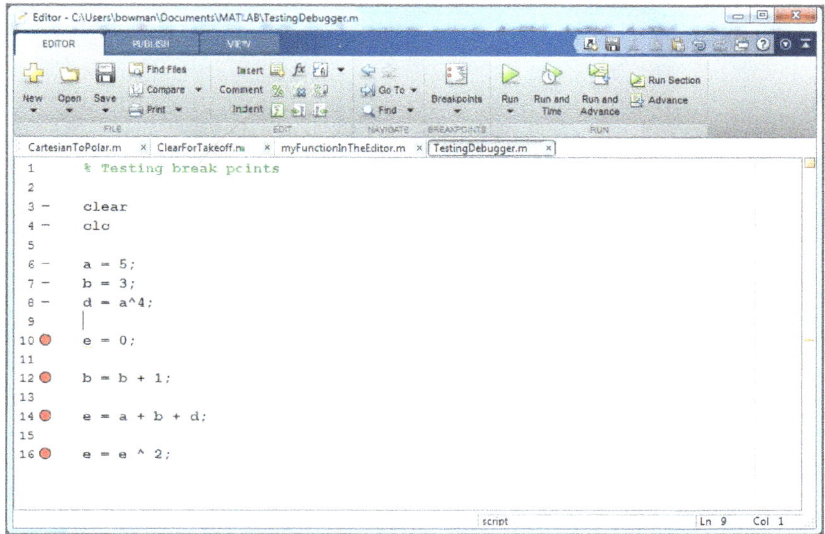

Stepping Through a Program

When using the Editor/Debugger, you must start by running the program or function. You can choose to step through the program one line at a time or have it continue until a breakpoint is encountered. Typically, you will set a breakpoint and execute the program and the Editor will automatically stop before the line the breakpoint is set on is executed. A green arrow will point to the program line when execution is paused. The controls in the Editor window will allow you to "step in" to functions you've written, or "step out" of functions you're currently in. Likewise, pressing the Continue button will resume execution—to completion or until the debugger encounters the next breakpoint.

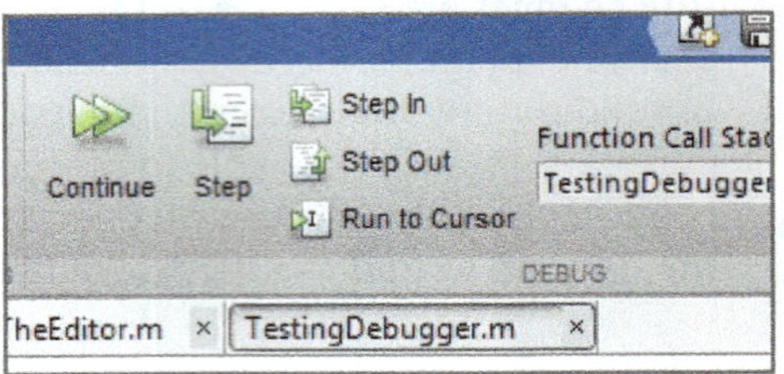

Examining Values

When the program is paused at a breakpoint or when you are stepping through the program a line at a time, you can view the value of any variable by hovering your mouse over the variable to see whether a line of code has produced the expected result. If the result is as expected, continue running or step to the next line. If the result is not as expected, then that line, or a previous line, contains an error. You can also examine the value of a variable in the Workspace window in the main MATLAB program window.

```
1       % Testing break points
2
3 -     clear
4 -     clc
5
6 -     a = 5;
7 -     b = 3;
8 -     d = a^4;
9           d: 1x1 double =
10 ●➪   e
11            625
12 ●    b = b + 1;
13
14 ●    e = a + b + d;
```

Correcting Problems and Ending Debugging

If a problem is difficult to diagnose, one method to help define or correct problems is to change the value of a variable in a paused program to see if continuing with the new value produces expected results. A new value can be assigned through the Command Window, Workspace browser, or Array Editor. Do not make changes to an M-file while MATLAB is in debug mode—it is best to exit debug mode before editing an M-file. If you edit an M-file while in debug mode, you can get unexpected results when you run the file.

In-Class Activities

ICA 17-1

Which of the following are not valid program/function filenames? Circle all that apply.

(A) `2b_solved.m`
(B) `calc_circum.m`
(C) `graph-data.m`
(D) `help4me.m`
(E) `MATLAB is fun.m`
(F) `matrix*matrix.m`
(G) `Mult2#s.m`
(H) `pi.m`
(I) `ReadFile.m`
(J) `SuperCaliFragiListicExpiAliDocious.m`

ICA 17-2

Without running these code segments in MATLAB, what will be the value stored in the specified variable after executing each program?

(a) What is stored in **`Dogs`**?

```
clear
clc
Greyhounds =30;
G_Cost =250;
Dalmations =10;
D_Cost =150;
Dogs =Greyhounds * G_Cost + Dalmations * D_Cost;
```

(b) What is stored in **`W`**?

```
clear
clc
X = 2;
Y = 4;
Z = 2;
W = (Y + Z/X + Z^1/2)
```

ICA 17-3

Without running these code segments in MATLAB, what will be the value stored in the specified variable after executing each program?

(a) What is stored in **`X`**?

```
clear
clc
A = 0;
B = 5;
X = A^B + B^A;
```

(b) What is stored in **Ramones**?

```
clear
clc
Joey = 1;
Johnny = 2;
DeeDee = 3;
Marky = 4;
Ramones = Joey * 2 + Johnny/2 + max([DeeDee, Marky]);
```

ICA 17-4

Write a program to store the following matrices into MATLAB and perform the following calculations. All variable assignments should be suppressed, but you should leave the calculations unsuppressed. If the variable assignment or calculation is not possible or causes an error message on the screen, write MATLAB comments explaining the problem.

Input variables:

$$A = \begin{bmatrix} 4 & 6 \\ -3 & 8 \end{bmatrix} \quad B = \begin{bmatrix} 3 & 5 \\ 5 & 5 \end{bmatrix} \quad C = [0 \quad 1] \quad D = \begin{bmatrix} 1 \\ 1 \end{bmatrix} \quad E = \begin{bmatrix} 3 & 6 & 5 \\ 7 & 1 & 5 \end{bmatrix} \quad F = \begin{bmatrix} 2 & 8 \\ 4 & 3 \\ 1 & 5 \end{bmatrix}$$

$$G = \begin{bmatrix} 7 & 8 & 3 \\ 4 & 8 & 2 \\ 5 & 9 & 5 \end{bmatrix} \quad H = [1 \quad 2 \quad \ldots \quad 299 \quad 300] \quad I = \begin{bmatrix} 200 \\ 198 \\ \vdots \\ -298 \\ -300 \end{bmatrix}$$

$$J = \begin{bmatrix} 20 & 40 & \ldots & 180 & 200 \\ 10 & 20 & \ldots & 90 & 100 \end{bmatrix} \quad K = \begin{bmatrix} 5 & 50 \\ 10 & 45 \\ \vdots & \vdots \\ 45 & -45 \\ 50 & -50 \end{bmatrix}$$

Calculations:

The result of the following calculations should each be stored into variable names of your choosing. Each calculation should be stored in a different variable. Here, note the "T" indicated the matrix should be transposed.

(a) A * B
(b) A + B
(c) A + C
(d) C − E
(d) E * F
(f) F * E
(g) G^2
(h) F^T
(i) C * D
(j) H + 15
(k) I * 30
(l) $E^T + 30$
(m) $E^2 * 50$
(n) $B^2 * 50$

ICA 17-5

Write a MATLAB program to evaluate the following mathematical expression. The equation should utilize variables for *a*, *b*, and *c*. Test the program with $a = 1$, $b = 2$, and $c = 3$.

$$x = \frac{b + \sqrt{b - 4a}}{c^7}$$

ICA 17-6

Write a MATLAB program to evaluate the following mathematical expression. The equation should utilize a variable for x. Test the program with $x = 30$.

$$A = \frac{x^2 \cos(2x + 1)}{(6x) \log(x)}$$

ICA 17-7

Write a MATLAB program to evaluate the following mathematical expression. The equation should utilize variables for x, μ, and σ. Test the program with $x = 0$, $\mu = 0$, and $\sigma = 1$.

$$P = \frac{1}{\sigma\sqrt{2\pi}} e^{-(x-\mu)^2/(2\sigma^2)}$$

This program calculates a *Gaussian normal distribution*.

ICA 17-8

Write a program that, given any two 2×2 arrays (`A` and `B`) as arguments, returns the sum of the two matrices. The program is not expected to work properly for any other matrix dimensions. You must calculate each term in `C` individually from terms in `A` and `B`—you may not use MATLAB's ability to add matrices directly. Use the following values to test the program: `A = [1 3; -2 2]` and `B = [-3 0; 4 -1]`.

ICA 17-9

Write a program that, given any two 2×2 arrays (`A` and `B`) as arguments, returns the matrix product. The program is not expected to work properly for any other matrix dimensions. You must calculate each term in `C` individually from terms in `A` and `B`—you may not use MATLAB's ability to multiply matrices directly. Use the following values to test the program: `A = [1 3; -2 2]` and `B = [-3 0; 4 -1]`.

ICA 17-10

The Shockley diode equation gives the relationship between the voltage (V) across a semiconductor junction and the current (`I`) through it.

$$I = I_0(e^{\left(\frac{V}{nV_T}\right)} - 1)$$

Assume V is a vector containing several voltage values. Write a MATLAB program that will calculate a vector `I` of the same length containing the current corresponding to each value in `V`.

Test your program with $I_0 = 2 \times 10^{-11}$, $n = 1$, $V_T = 25.85 \times 10^{-3}$, and $V = [0.4, 0.55, 0.65, -5,0]$.

ICA 17-11

For each MATLAB code segment shown, write the function header necessary to convert the code segment into a function. Use the guidelines specified with each problem.

(a) Assume there only needs to be one function output, the variable **`X`** and that the function will be stored in a file named **`CandyCrush.m`**.

```
X = (A + B + B^A)*pi;
```

(b) Assume the function should have five output variables: the paintable area of each wall (all 4 walls) and the total paintable area. You may assume this function will be stored in a file named **FullHouse.m**.

```
P1 = Wall1 * RoomHeight;
P2 = Wall2 * RoomHeight;
P3 = Wall3 * RoomHeight;
P4 = Wall4 * RoomHeight;
PaintableArea = P1 + P2 + P3 + P4;
```

(c) Assume the function should have 3 output variables: the minimum value, the maximum value, and the average value. You may assume this function will be stored in a file named **MinMaxMean.m**.

```
MyMin = min([D,D^2,D*10]);
MyMax = max([D^3,D*3,D + 3000]);
MyMean = mean([D/2,D/3,D/4]);
```

ICA 17-12

Assume you are given the following function:

```
function [status] = ClearForTakeoff(T)
status = [T +2, T -4; T*6, T/8];
```

(a) What is stored in the variable Z after executing the following lines of MATLAB code?

```
A = 2;
Z = ClearForTakeoff(A);
```

(b) What is stored in the variable E after executing the following lines of MATLAB code?

```
D = 0;
E = ClearForTakeoff (D);
```

ICA 17-13

A member of your team gives you the following MATLAB program. Your job is to create the functions used in the program.

```
A = 10; % area [cm^2]
H = 30; % height [cm]
V = 50; % volume [cm^3]
% calculates the radius [cm] of the circle
R1 = RadiusCircle(A);
% calculates the radius [cm] of a cone
R2 = RadiusCone(H,V);
```

ICA 17-14

Assume you are given the following function:

```
function [M_NEW,N] = SimpleCalculations(Wii)
M_NEW = Wii + 30;
N = M_NEW/2;
Wii = 35;
```

What is stored in the variables *P*, *C*, and *r* after executing the following lines of MATLAB code?

```
r = 5;
[P,C] = SimpleCalculations(r);
```

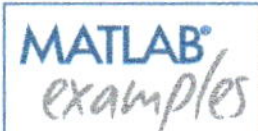

ICA 17-15

A novice programmer has attempted to write a program and two functions using global variables. These three files are provided for you with the online materials.

The purpose of the code is to determine the total pressure at the bottom of a cylindrical container filled with a liquid on an arbitrary planet, and to determine the potential energy of that cylinder at a given height above the planet's surface.

(a) Without running the code, determine what the values of the following variables will be following execution of the code: **`g, atm, Radius, Depth, CylMass, rho, Height, H, Pressure, P, CylVol, ContentsMass, TotalMass, PE`**

(b) Which of the variables in part (a) DO NOT appear in the main program workspace?

(c) For the values specified in the program, the correct answers are `Pressure` = 26,150 [Pa] and `PE` = 19,987 [J]. The program and associated functions calculate the pressure correctly, but the potential energy is wrong. Explain what went wrong and how to repair the problem.

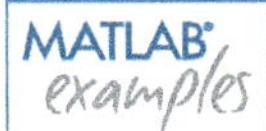

ICA 17-16

Debug the MATLAB programs/functions provided for you with the online materials.

```
atm to Pa.m
ft to m.m
ICA #7.m
```

These files must be corrected to eliminate any syntax, runtime, formula coding, and formula derivation errors. You may assume that the header comments at the top of the program and functions are correct. In addition to the corrected files, you must submit an algorithm template, showing the main program and both functions.

ICA 17-17

Debug the MATLAB programs/functions provided for you with the online materials.

```
SG to rho.m
N_2_lbf.m
ICA S-D.m
```

These files must be corrected to eliminate any syntax, runtime, formula coding, and formula derivation errors. You may assume that the header comments at the top of the program and functions are correct. In addition to the corrected files, you must submit an algorithm template, showing the main program and both functions.

ICA 17-18

Consider the following MATLAB program and function, stored in MATLAB's Current Directory:

MyRadFunction.m

```
clear clc
function [Out] = RadF (In1, In2, In3)
Out = (In1+In2)/2 + (In2+In3)/4;
```

MyRadProgram.m

```
clear clc
InVar1=1; InVar2=3; InVar3=-1;
M=MyRadFunction(invar1), MyRadFunction(invar2),
myradfunction(invar3);
```

Fix the program and function to eliminate all of the error messages. Note that for the variables provided in the program as `InVar1`, `InVar2`, and `InVar3`, the numerical result stored in `M` should be 2.5.

ICA 17-19

A novice MATLAB user created the following code with poor choice of variable names. Correct the errors and fill in the blanks provided to comment this code to determine the purpose of the code.

Main Program	Comments
`T = [30, 45, 120, 150];`	**(a)** `% T =`
`Z = 2;`	**(b)** `% Z =`
	`% The purpose of the function DTOR is ...`
`[W] = DTOR(T)`	**(c)** `%`
	(d) `% W =`
	`% The purpose of the function PCAR is ...`
	(e) `%`
`[X,Y] = PCAR(Z,W)`	**(f)** `% X =`
	(g) `% Y =`

```
function [A] = DTOR(T)
A = T*2*pi/360;
```

```
function (P,Q) = PCAR(M,N)
P = M * cos(N);
Q = M * sin(N);
```

(h) The output of this code, when run, is ______

Chapter 17 REVIEW QUESTIONS

1. The specific gravity of gold is 19.3. Write a MATLAB program that will determine the length of one side of a 0.4 kilogram cube of solid gold, in units of inches.

2. An unmanned X-43A scramjet test vehicle has achieved a maximum speed of Mach number 9.68 in a test flight over the Pacific Ocean. Mach number is defined as the speed of an object divided by the speed of sound. Assuming the speed of sound is 343 meters per second, write a MATLAB program to determine the record speed in units of miles per hour.

3. A rod on the surface of Jupiter's moon Callisto has a volume of 0.3 cubic meters. Write a MATLAB program that will determine the weight of the rod in units of pounds-force. The specific gravity is 4.7. Gravitational acceleration on Callisto is 1.25 meters per second squared.

4. The Eco-Marathon is an annual competition sponsored by Shell Oil, in which participants build special vehicles to achieve the highest possible fuel efficiency. The Eco-Marathon is held around the world with events in the United Kingdom, Finland, France, Holland, Japan, and the United States.

   A world record was set in the Eco-Marathon by a French team in 2003 called Microjoule with a performance of 10,705 miles per gallon. The Microjoule runs on ethanol. Write a MATLAB program to determine how far the Microjoule will travel in kilometers given a user-specified amount of ethanol, provided in units of grams. For your test case, you may assume that the user provides 100 grams of ethanol.

5. Write a program to determine the mass of oxygen gas (formula: O_2, molecular weight = 32 grams per mole) in units of grams in a container. You may assume that the user will provide the volume of the container in units of gallons, the temperature in the container in degrees Celsius, and the pressure in the container in units of atmospheres. For your test case, you may assume that the user provides 1.25 gallons for the volume of the container, 125 degrees Celsius for the temperature, and 2.5 atmospheres for the pressure in the container.

6. Write a program to calculate a temperature provided by the user in units of Fahrenheit to units of kelvins. As a test case, you may assume the user provides the temperature of −129 degrees Fahrenheit, which is the world's lowest recorded temperature.

7. Write a program to determine how long, in units of seconds, it will take a motor to raise a load into the air. Assume the user will specify the power of the motor in units of watts, the rated efficiency as a percentage (in whole number form—for example, 50 for 50%), the mass of the load in kilograms, and the height the load is raised in the air in units of meters. As a test case, you may assume the user provides the 100 watts for the power of the motor, 60% for the efficiency of the motor, 100 kilograms for the mass of the load, and 5 meters for the height the load is raised.

8. A cylindrical tank filled to a height of 25 feet with tribromoethylene has been pressurized to 3 atmospheres ($P_{surface}$ = 3 atmospheres). The total pressure in at the bottom of the tank is 5 atmospheres. Write a MATLAB program to determine the density of tribromoethylene in units of kilograms per cubic meter.

9. Write a MATLAB program that implements the quadratic equation. Recall the quadratic equation:

$$r = \frac{-b \pm \sqrt{b^2 - 4ac}}{2a}$$

10. Write a MATLAB program that implements the Pythagorean theorem. Recall that the theorem states that the length of the hypotenuse (z) can be calculated by the sum of the squares of the adjacent sides (x and y), or

$$z = \sqrt{x^2 + y^2}$$

11. The specific gravity of gold is 19.3. Write a MATLAB function that will determine the length of one side of a cube of solid gold, in units of inches, provided the mass of the cube in kilograms.

12. An unmanned X-43A scramjet test vehicle has achieved a maximum speed of Mach number 9.68 in a test flight over the Pacific Ocean. Mach number is defined as the speed of an object divided by the speed of sound. Assuming the speed of sound is 343 meters per second, write a MATLAB function that will calculate the speed in units of miles per hour given the Mach number.

13. A rod on the surface of Jupiter's moon Callisto has a volume of 0.3 cubic meters. Write a MATLAB function that will determine the weight of the rod in units of pounds-force given the rod volume in cubic meters. The specific gravity is 4.7. Gravitational acceleration on Callisto is 1.25 meters per second squared.

14. A cylindrical tank filled to a height of 25 feet with tribromoethylene has been pressurized to 3 atmospheres ($P_{surface}$ = 3 atmospheres). The total pressure in at the bottom of the tank is 5 atmospheres. Write a MATLAB function that will determine the density of tribromoethylene in units of kilograms per cubic meter given the height in units of feet and the surface and total pressures in units of atmospheres.

15. Write a function that implements the quadratic equation. Given three inputs (`a`, `b`, and `c`), calculate the roots (`r1` and `r2`) of the quadratic formula. Recall the quadratic equation:

$$r = \frac{-b \pm \sqrt{b^2 - 4ac}}{2a}$$

16. Write a function that implements the Pythagorean theorem. Recall that the theorem states that the length of the hypotenuse (`z`) can be calculated by the sum of the squares of the adjacent sides (`x` and `y`), or

$$z = \sqrt{x^2 + y^2}$$

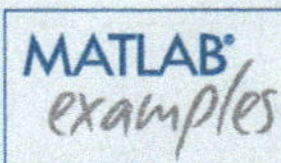

17. In the starting file provided on online, there are data sets from two different data collection sessions. In the first data collection session, a lab technician collected three different measurements and recorded mass in grams, height in feet, and time in minutes. However, in data collection session two, a different lab technician collected four different measurements and recorded mass in pounds-mass, height in centimeters, and time in hours.

Your job is to write a function that will calculate the potential energy in joules and power in watts for each data condition (7 total). In addition, you will need to write a program to call the function using the datasets provided in the starting file. Your function should only consider variables in SI units, so you will need to convert the vectors in the program before passing them in to your function. Note that all conversions must be done in MATLAB code—you may not hard code any values you calculate by hand.

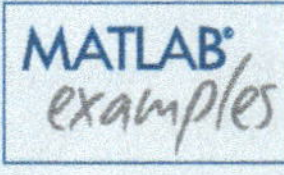

18. As part of a team investigating the effect of mass on the oscillation frequency of a spring, you obtain data from three different lab technicians, provided for you in the starter file online.

The data consists of frequency data on three different springs recording the amount of time it takes each spring with different masses attached to oscillate a certain number of times. In this experiment, each technician recorded the time it took for the spring to oscillate 25 times, stored in the variable *N* in the starter file.

As part of the analysis, you need to write a program containing the experimental data provided. The data should be converted to make all units consistent. The data should then be passed into a function.

The function should accept three inputs: a vector containing mass measurements (grams), a vector containing time measurements (seconds), and a variable containing the number of oscillations observed in the experiment.

The function should return two matrices. The first matrix should contain the mass measurements (kilograms) in the first column and the period (seconds), or the length of time required for one oscillation for each mass in the second column. The second matrix should contain a calculation of the force applied (newtons) by each mass in the first column and the frequency, the number of oscillations per second, or the inverse of the period (hertz) for each mass measured in the second column.

19. We have made many measurements of coffee cooling in a ceramic coffee cup. We realize that as the coffee cools, it gradually reaches room temperature. Consequently, we report the value of the coffee temperature in degrees above room temperature (so after a long time, the temperature rise will be equal to 0). Also, we realize that the hotter the coffee is initially (above room temperature), the longer it will take to cool. The values presented here are in degrees Fahrenheit.

Temperature Rise (T) [°F]

Initial Temp Rise (T_0) [°F]	Cooling Time Elapsed (t) [min]					
	0	10	20	30	40	50
20	20	13	9	6	4	3
40	40	27	18	12	8	5
60	60	40	27	18	12	8
80	80	54	36	24	16	11

Write a MATLAB function that will perform a single interpolation given five numbers as input arguments and return the interpolated value as the only function output.

Write a MATLAB program that will calculate the following scenarios. Store each part in a different variable (e.g., part **(a)** should be stored in a variable named `PartA`, part **(b)** should be stored in a variable named `PartB`, etc.).

(a) What is the temperature (rise) of the cup of coffee after 37 minutes if the initial rise of temperature is 40 degrees Fahrenheit?

(b) If the coffee cools for 30 minutes and has risen 14 degrees Fahrenheit at that time, what was the initial temperature rise?

(c) Find the temperature rise of the coffee at 17 minutes if the initial rise is 53 degrees Fahrenheit.

20. In a factory, various metal pieces are forged and then plunged into a cool liquid to quickly cool the metal. The types of metals produced, as well as their specific heat capacity, are listed in the table below.

Material	Specific Heat [J/(g °C)]
Aluminum	0.897
Cadmium	0.231
Iron	0.450
Tungsten	0.134

The metal pieces vary in mass, and are produced at a temperature of 300 degrees Celsius. The ideal process lowers the temperature of the material to 50 degrees Celsius. The liquid used to cool the metal is glycerol. The properties of glycerol are listed below.

Material Property	Value [Units]
Specific heat	2.4 J/(g °C)
Specific gravity	1.261
Initial temperature	25 °C

There are data sets from four different data collection sessions. In the first data collection session, a lab technician collected seven different measurements and recorded mass in grams of aluminum rods. The second data set contains cadmium rods; the third data set contains iron rods, and the fourth data set contains tungsten rods.

Mass of Object [g]			
Aluminum	Cadmium	Iron	Tungsten
2,000	3,000	2,500	4,800
2,500	4,000	3,500	6,400
3,000	6,500	4,500	10,400
4,000	8,000	5,000	12,800
5,500	10,000	5,500	16,000
7,500	11,000	7,500	17,600
8,000	15,000	9,000	24,000

Your job is to write two functions: (1) to calculate the thermal energy in joules that must be removed for each rod to cool it from 300 to 50 degrees Celsius for each mass and (2) to determine the volume of fluid needed in gallons to properly cool the rod for each mass. The result from the second function should be a matrix with the first column being the mass of the rod, and the second column the volume of fluid needed.

In addition, you will need to write a program to call the functions using the datasets provided in the final table. You will call each function four times, once for each material. Your function should only consider variables in SI units, so you will need to convert the vectors in the program before passing them to your function as necessary. Note that all conversions must be done in MATLAB code—you may not hard code any values you calculate by hand.

21. Assume you are sent a file name `VolumeOfBox.m`, which is a function that calculates the volume of a box, given the length (`L`), width (`W`), and height (`H`) in non-specific units. The assumption of the function provided to you is that the length, width, and height variables you provide the function are all measured in the same unit. When you attempt to call the MATLAB function, you notice that it crashes with the following error message:

```
Error using VolumeOfBox
Too many input arguments.
```

The contents of the `VolumeOfBox.m` file are shown below:

```
function [L,W,H] = VolumeOfBox(V)
% Calculate the volume of a box
V = L * W * H;
```

(a) In your debugging, you have discovered that the error message on the screen means that the function header is incorrect. Write the correct function header.

(b) Write a line of MATLAB code that will call the `VolumeOfBox` function with the value 3 for the length, 5 for the width, and 7 for the height variables and store the volume calculation in a variable named `BoxVol`.

22. Assume you are sent a file named `GetCostByTime.m`, which is a function that calculates two different cost calculations given a time variable (`t`, non-unit specific) and a scaling factor, `S`. In the source code, the cost calculations are stored in two variables, `M` and `N`. When you attempt to call the MATLAB function, you notice that it crashes with the following error message:

```
Attempt to execute SCRIPT GetCostByTime as a function
```

The contents of the `GetCostByTime.m` file are shown below:

```
t = t/s;
% Calculate the total cost
M = t^2 + 3*t + 10;
N = t^2 - 2*t + 30;
```

(a) In your debugging, you have discovered that the function header is missing from this file. Write the correct function header for the `GetCostByTime.m` file.

(b) Write a line of MATLAB code that will call the `GetCostByTime` function with the value 360 for the time variable and 60 as the scaling factor and store the two different cost calculations in variables `C1` and `C2`.

23. You have been assigned to a new project at work. The previous engineer had created the following file, which your new boss claims does not work. Debug the program to correct all errors. This file must be corrected to eliminate any syntax, runtime, formula coding, and formula derivation errors. You may assume that the header comments at the top of the program and all comments throughout the program are correct.

```
% Problem Statement: A ball is thrown vertically into the air with an
% initial kinetic energy of 2,500 joules. As the ball rises, it gradually
% loses kinetic energy as its potential energy increases. At the top of
% its flight, when its vertical speed goes to zero, all of the kinetic
% energy has been converted into potential energy. Assume that no energy
% is lost to frictional drag, etc. How high does the ball rise in units of
% meters if it has a mass of 5 kilograms?
% Input Variables
% KE - Kinetic Energy [J]
% m - Mass of ball [kg]
% Output Variables
% H - Max height of ball [m]
% Other Variables
% g - Acceleration due to gravity [m/s^2]
% Assumptions
% No energy losses due to friction
% g = 9.8 m/s^2

clearscreen
clearworkspace

% Set initial variables
2500 = KE;
m = 5 kg;
gravity = 9.8;

% Calculate height, assuming Kinetic Energy = height * mass * gravity
KE(2500) = H * m * g
```

24. You have been assigned to a new project at work. The previous engineer had created the following files, which your new boss claims do not work. Debug the program to correct all errors.

These files must be corrected to eliminate any syntax, runtime, formula coding, and formula derivation errors. You may assume that the header comments at the top of the program and all comments throughout the program are correct.

```
% Problem Statement: A cylindrical tank filled to a height of 25 feet with
% tribromoethylene has been pressurized to 3 atmospheres (P_s = 3 atm).
% The total pressure (P_t) at the bottom of the tank is 5 atmospheres.
%
% This program will determine the density of the tribromoethylene
% in units of kilograms per cubic meter.
%
% Inputs:
%       Height              H     25  ft
%       Surface Pressure    P_s   3   atm
%       Total Pressure      P_t   5   atm
% Outputs:
%       Density             rho   ?   kg/m^3
```

```
% Assumptions:
%       Gravity                    g       9.8 m/s^2
% Define input variables
25 ft = H;
3 atm = P_s;
5 atm = P_t;
9.8 m/s^2 = g;

% Convert total and surface pressure from atm to Pa
Psurf = P*1,211,458;
P_t = P_t*14.7;

% Convert height from ft to m
H*3.28;

% Calculate density, knowing the calculation of total pressure using equation
% P_t = P_s + (rho)(g)(H)
P_t = P_s + (rho)(g)(H)
```

25. You have been assigned to a new project at work. The previous engineer had created the following files, which your new boss claims do not work. Debug the program to correct all errors.

These files must be corrected to eliminate any syntax, runtime, formula coding, and formula derivation errors. You may assume that the header comments at the top of the program and all comments throughout the program are correct.

```
% Problem Statement: A rod on the surface of Jupiter's moon Callisto has a volume of
% 0.3 cubic meters and specific gravity of 4.7. The gravitational constant is 1.25
% meters per second squared. Determine the weight of the rod in units of
% pounds-force.
%
% Inputs:
%       Specific Gravity      SG      4.7 [-]
%       Volume                V       0.3 m^3
% Outputs:
%       Weight                w       ??? lbf
% Other variables:
%       Density               rho     ??? kg/m^3
% Assumptions:
%       Gravity               g       1.25 m/s^2
%       Density of water      r_w     1000 kg/m^3

% Define input variables
25 = SG;
g = 1.25 m/s^2;
RHO_water = 1;
VOL = 0.3 m^3;

% Calculate rod density [kg/m^3] using specific gravity
SG=Rho*r_w/1000*62.4;

% Calculate weight of rod
w = (Rho / V) (g);
```

CHAPTER 18
INPUT/OUTPUT IN MATLAB

When writing a program, it is wise to make the program understandable and interactive. This chapter describes the mechanisms in MATLAB that allow the user of a program to input values during runtime. In addition, this chapter describes a few different mechanisms for displaying clean, formatted output to the user so that the results can be read and interpreted.

18.1 INPUT

LEARN TO:

Accept numeric and character input with the input function
Debug input statements when error messages appear on the screen
Create menus to allow interactive input

If all programs required the programmer to enter the input quantities at the beginning of the program, everyone would need to be a computer programmer to use the program. Instead, MATLAB and other programming languages have ways of collecting input from the program's user at runtime. This section focuses on program-directed user input, using three different approaches.

Numerical Input Typed by the User

The **input** function allows the user to input numerical data into MATLAB's Command Window. The user is prompted to input values at the point the input statement occurs in the code.

```
Variable = input ('String')
```

- ***Variable:*** The variable where the input value will be stored
- ***'String':*** The text that will prompt the user to type a value

It is important to note the difference between user input and functional input. User input prompts the user in runtime to type in a value. Functional input are values passed in as arguments to a function before a function is executed. Normally, all user input occurs in programs instead of functions.

EXAMPLE 18-1

Imagine you are writing a program to calculate the speed of a traveling rocket. Prompt the user to input the distance the rocket has traveled.

Before writing an input statement, it is critical to remember that MATLAB cannot handle units, so it is up to the programmer to perform any unit conversions necessary in the algorithm. Imagine that the user wants to report the speed of the rocket in miles per hour. It would be wise to have the user input the distance traveled in miles and the travel time in hours.

```
Distance = input('How far has the rocket traveled [miles]?')
Time = input('How long has the rocket been in the air? [hours]')
Speed = Distance/Time
```

In the preceding code segment, the functions were executed without a semicolon being at the end of the input call. If you insert a semicolon at the end of the input call, output of the stored variables to the screen is suppressed, which allows for cleaner display on the Command Window.

NOTE

The use of a semicolon (;) suppresses output to the Command Window.

```
% Variable output suppressed
Distance = input('How far has the rocket traveled [miles]?');
Time = input('How long has the rocket been in the air [hours]?');
```

COMPREHENSION CHECK 18-1

(a) Write an input statement to ask for the user's height in inches.
(b) Write an input statement to ask the user to enter the temperature in degrees Fahrenheit.

Text Input Typed by the User

The input function also allows the user to input text data into MATLAB's Command Window. The programmer must include an additional argument (`'s'`) to the `input` function to tell MATLAB to interpret the user input as text. The term `'s'` is used to indicate a string.

Variable = input ('String','s')

- ***Variable:*** The variable where the input value will be stored
- ***'String':*** The text that will prompt the user to type a value
- ***'s':*** Input type is text, not numeric

EXAMPLE 18-2

Imagine you are required to ask for the user's name and for the name of the month he/she was born. Suppress all extraneous output.

```
Name = input('Type your full name and press enter: ','s');
DataMonth = input('What month were you born? ','s');
```

COMPREHENSION CHECK 18-2

(a) Write an input statement to ask the user for the color of his/her eyes.
(b) Write an input statement to ask the user to type the current month.

Debugging Input Statements

In the next chapter, you will learn how you can use conditional statements to prevent errors from occuring on the screen due to improper input from the user, but there are some errors that may appear during execution that will impact the flow of your program or function. Consider a simple line of MATLAB code that expects the user to type in a number:

```
M = input ('Type a number:')
```

The correct user behavior when this line of code is executed is to type a number and press enter, however, this is not always the case. Sometimes the user will accidentally include a typo in their input, or in nastier situations, the code might be facing a malicious user attempting to break the code for a variety of reasons. Assume you type the input statement in a program. If the user types the number 4 and presses the `Enter` key, the variable `M` contains the number 4, as expected.

```
Type a number: 4
M =
   4
```

However, what happens when the user types in a word or phrase? There are a few different error messages that may appear on the screen when improper text is provided to the input function. Consider the following error message sequence:

```
Type a number: there are four cats outside
Error: Unexpected MATLAB expression.

Type a number: 4cats
 4cats
   |
Error: Unexpected MATLAB expression.
>>
```

On the user's first attempt, MATLAB simply did not recognize the sequence of nonnumeric characters, issued the error message, and redisplayed the prompt for the user to try again. On the second attempt, the user began with a digit, making MATLAB think that the input was valid. However, when the MATLAB input interpreter saw an alphabetic character next, MATLAB became completely confused, issued the error message, terminated the program, and returned to the command prompt. Consider this error message and program termination as MATLAB's way of saying "I have no idea what's going on." If MATLAB has some idea where the error occurs, it will repeat the user provided expression and display the pipe (|) character below the character where it thinks the first invalid character appears.

Consider the following error message sequence:

```
Type a number: four
Error using input
Undefined function or variable 'four'.

Type a number: four+cats
Error using input
Undefined function or variable 'four'.
```

```
Type a number: [4; cats]
Error using input
Undefined function or variable 'cats'.

Type a number: Cats(4)
Error using input
Undefined function 'Cats' for input arguments of type 'double'.
```

In these error messages, the "**Undefined function or variable**" error message indicates that MATLAB is attempting to use the user input as a MATLAB expression and treat some values as variables or function names. In the error messages above, the word in single quotes is the value MATLAB is attempting to interpret as either a function or variable name. None of these input attempts results in program termination, and the user is prompted again for input.

Additionally, the `input` function will attempt to evaluate an expression typed by the user if there are valid variables referenced in the user input value. Assume the variable `rats` is defined in MATLAB's workspace and contains the value 10:

```
Type a number: rats+5
M =
    15
```

This expression did not result in an error message because the `rats` variable existed in MATLAB's workspace, so MATLAB evaluated the expression and stored the result in the variable `M`.

```
Type a number: Cats = four
Error: The expression to the left of the equals sign is not a valid
target for an assignment.
```

In some error messages, MATLAB may report one problem, but fail to report other issues. In the error message above, MATLAB is complaining about the expression to the left of the equals sign not being a valid target for an assignment. However, this error message fails to point out that the "`Cats`" and "`four`" are not variables in MATLAB's workspace. As soon as MATLAB discovers the first problem, it quits looking for additional issues and displays the error on the screen. Removing the equals sign from the input will result in a different error message.

While some values typed as input into MATLAB when the interpreter is expecting numeric input will result in a nasty error message, sometimes no error or warning messages are generated and the program will continue as if it has valid data. Consider the following scenario, using the same `input` expression, where MATLAB is expecting a number, but the user types a word as a string (enclosed by single quotes):

```
Type a number: 'cats'
M =
cats
```

Note that the result of this action is the string `'cats'` being successfully stored in the variable `M`. This is equivalent to using the `input` function with the extra argument denoting that MATLAB should expect character input, but that method does not require the user to type the input enclosed by single quotes. In the next chapter, you will see how functions like `isnumeric` or `ischar` can be used to determine if a variable is of a particular type, which will be necessary for assuring your program uses valid input of the correct data type.

In addition to allowing the user to type numeric input or a string, MATLAB will also allow the user to provide no input at all. Consider the following scenario, where the user only presses the `Enter` key without typing any other text after the input statement:

```
Type a number:        (no text, the user only pressed Enter)
M =
    []
```

In this situation, the value stored in the variable `M` is an empty vector, which is different from an empty string or the number zero. The empty vector in MATLAB is an absence of any data and is similar in nature to the concept of having a null value, which is common in other programming languages like C.

Note that all of the scenarios presented here discussed the input statement when it was expecting numeric input. The string input statement will not produce as many errors because the input provided by the user is written to a variable as a sequence of characters verbatim. Even in the scenario where the user presses the `Enter` key without typing any information, the result is an empty string.

```
M = input ('Type a number:','s')
Type any word: cats
M =
cats

Type any word:        (no text,the user only pressed Enter)
M =
    ''

Type any word: 4
M =
4
```

In this final situation, the value 4 is stored in the variable `M` as a number, even though the input requested a text string. The numerical value could be used in a calculation without an error occurring, or it can be used as a text string.

Menu-Driven Input

In the preceding example, the user was required to type the name of the month he/she were born, but there is no good way to predict the way the user will actually type the name of the month into MATLAB. For example, the user may abbreviate, represent the number numerically, or misspell the name of the month completely. MATLAB has a built-in input function (**menu**) that allows the programmer to specify a list of options for user selection. The menu function will display a graphical prompt of a question and responses.

Variable = menu ('String', 'opt1','opt2',...)

- ***Variable:*** The variable where the ordinal value will be stored
- ***'String':*** The text that will prompt the user to select an item in the menu
- ***'optN':*** The options that appear as buttons in the menu

The first argument of the menu function is always the question to display (`'String'`), and the following arguments are the possible responses. The menu function does not store the actual text response, but rather the ordinal number of the response in the list. An example will help clarify this idea.

EXAMPLE 18-3

Suppose we want to ask the user to input a favorite color. However, we want to force the user to choose from red, green, or blue. Suppress all extraneous output.

```
Color = menu('Favorite Color','Red','Green','Blue');
```

This command will create the menu shown. The value returned upon selection is an integer. For example, if the user were to select "Red" in the menu, the value stored in `Color` *would be 1 since "Red" is the first option available in the list.*

COMPREHENSION CHECK 18-3

(a) Create a menu to ask the user to select the current month.

(b) Given the menu statement below, what is stored in variable `C` when the user clicks "Red"?

```
C = menu('My favorite color:','Green','Blue','Yellow',
'Black','Red')
```

In addition to the `menu` function, the **questdlg** function will allow you to create a pop up dialog that allows the user to respond by clicking one of up to three different options for answers. The default buttons are yes, no, and cancel, although these can be customized by the programmer. The `questdlg` function returns the text shown on the button clicked, not the ordinal value.

Variable = questdlg ('qstring', 'title', 'opt1', 'opt2', 'opt3')

- ***Variable:*** The variable where the ordinal value will be stored
- ***'qstring':*** The question being asked
- ***'title':*** Creates a title in the dialog box title bar
- ***'optN':*** The options that appear as buttons. If nothing is listed, the options will be yes, no, and cancel.

Consider the following example:

```
A = questdlg ('My favorite color is green', 'Color')
```

If the user chooses the No button, then the word "No" will be stored in the variable `A`.

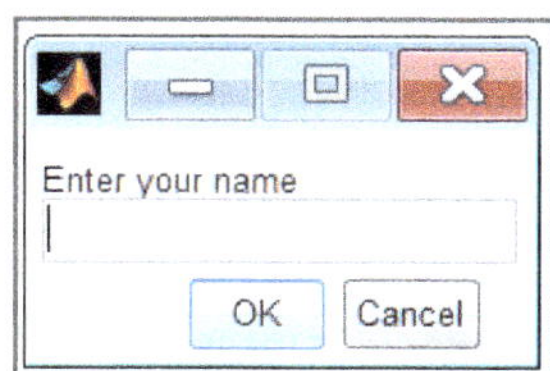

In the example below, the dialog box shown is created.

```
A = questdlg ('My favorite season is…', 'Season','Winter',
'Summer','Other')
```

Similarly, the **inputdlg** function will display a graphical input box to the user to allow them to type a string into boxes rather than typing their responses into the Command Window. Consider the following example:

```
K = inputdlg('Enter your name')
```

For more information on these functions, refer to MATLAB's documentation on the `questdlg` and `inputdlg` functions.

18.2 OUTPUT

LEARN TO:

Display the contents of a variable using `disp`
Create formatted display (in sentences) using `fprintf`
Save formatted output to a string using `sprintf`

As critical as input is to enhancing the versatility of a program, output is equally important—this is how the user gets feedback on the computer's solution. Furthermore, as discussed earlier, program output of intermediate calculations is an important diagnostic tool. Once a program is working correctly, you should eliminate superfluous program output by using the semicolon, as demonstrated earlier. The program user will also find it helpful if the remaining output is neatly formatted.

Using `disp` for User Output

One of the basic methods for displaying the contents of any variable is through the use of the **`disp`** function.

```
disp (Variable)
```

- ***Variable:*** The variable to be displayed on the screen

Given some variable *x* defined in MATLAB's Workspace, `disp(x)` displays the value of *x* to the screen. However, the `disp` command cannot display a value and text on the same line. Furthermore, `disp` relies on using MATLAB's default numerical display format for displaying any numbers, so it does not provide a capability for changing the form in which the number is displayed.

Formatting Output with `fprintf`

The formatted print command, **fprintf**, gives extensive control over the output format, including spacing.

```
fprintf ('String', Var1, Var2, . . .)
```

- ***'String':*** The formatted output string
- ***VarN:*** The variables or values to be inserted into the formatted output string

The `fprintf` command can intersperse precisely formatted numbers within text, guarantee table alignment, align the decimal points of a column of numbers, etc. The `fprintf` command uses control and format codes to achieve this level of precision, as shown in Tables 18-1 and 18-2.

Table 18-1 Output control with `fprintf`

Control Character	Use	Example
\n	Inserts a new line	`fprintf('Hello\n');`
\t	Inserts a tab	`fprintf('A\tB\tC\n');`
\\	Inserts a backslash	`fprintf('Hello\\World\n');`
'' (two single quotes)	Inserts a single quote mark	`fprintf('Bob''s car\n');`
%%	Inserts a percent symbol	`fprintf('25%%');`

The biggest benefit of using `fprintf` is the ability to plug values contained in variables into formatted sentences. In addition to the formatting controls shown above, MATLAB contains a number of special controls to represent different types of variables that might be displayed within a sentence. Inserting a variable control character within a string tells MATLAB to plug in a variable at that location in the string. Note that MATLAB's documentation on `fprintf` contains additional control characters not covered in this textbook that you may find useful in your programming.

Table 18-2 Variable (numeric) and TextVariable (text) control with `fprintf`

Control Character	Use	Example: Variable = 13/15;
%f	Inserts a value shown to 6 decimal places	`fprintf('%f', Variable)` `0.866667 is displayed`
%0.Mf	Inserts a fixed point value with M decimal places	`fprintf('%0.1f',Variable);` `0.9 is displayed`
%N.Mf	Inserts a value with a field width of N (left padded with spaces), containing M decimal places	`fprintf('%7.2f',Variable)` `_ _ _ 0.87 is displayed, where _ indicates a blank space`
%0N.Mf	Inserts a value with a field width of N (left padded with zeros), containing M decimal places	`fprintf('%07.2f',Variable)` `0000.87 is displayed`
%e or %E	Inserts a number in exponential notation	`fprintf('%e',Variable);` `8.66667e-01 is displayed` `fprintf('%0.3E',Variable);` `8.667E-01 will display`
%s	Inserts text	`fprintf('%s',TextVariable);`

EXAMPLE 18-4

Assume that the following variables are defined in MATLAB's Workspace:

```
Age=20;
Average=79.939;
Food='pizza';
Postal=515;
```

Display the variables in formatted `fprintf` statements. At the end of each formatted output statement, insert a new line.

Since age is a decimal value, it is best to use the `%0.0f` *control character:*

```
fprintf('My age is %0.0f.\n',Age);
```

To display the average with two decimal places:

```
fprintf('The average is %0.2f.\n',Average);
```

Note that MATLAB rounds values when using `%f`.

To display a string within a sentence:

```
fprintf('My favorite food is %s.',Food);
```

To display a postal code, the field width of a postal code is typically five numbers long:

```
fprintf('My postal code is %05.0f.\n',Postal);
```

EXAMPLE 18-5

Assume the variables in Example 18-4 are still defined in the workspace. Create a single formatted output statement that displays each sentence with a line break between each sentence.

```
fprintf('My age is %0.0f.\n The average is %0.2f.\n
My favorite food is %s.\n My postal code is %05.0f.\n',
Age,Average,Food,Postal);
```

In MATLAB, the previous line would be typed on a single line. Note that the order of the arguments to the function corresponds to the order of the variables in the string.

COMPREHENSION CHECK 18-4

Assume that the variable `M` is stored in the workspace with the value 0.3539.

(a) What would be the control code used to display `M` with two decimal places?
(b) What would be the control code used to display `M` with three decimal places in scientific notation?
(c) What would be the output of `M` if a control code is used to display it to two decimal places?
(d) What would be the output of `M` if a control code is used to display it to three decimal places in scientific notation?
(e) What would be the output of `M` if a control code is used to pad the field width to 10 characters, using zero padding, and showing 3 decimal places?

COMPREHENSION CHECK 18-5

Consider the following segment of code. What appears in the Command Window if this is executed?

```
clear
clc
S=[2012,27,17;2011,34,13;2010,29,7];
fprintf('Clemson-USC Football Rivalry:\n\n');
fprintf('Year\tUSC\t\tClemson\n');
fprintf('%0.0f\t%0.0f\t\t%0.0f\n',S(1,1),S(1,2),S(1,3));
fprintf('%0.0f\t%0.0f\t\t%0.0f\n',S(2,1),S(2,2),S(2,3));
fprintf('%0.0f\t%0.0f\t\t%0.0f\n',S(3,1),S(3,2),S(3,3));
```

Saving formatted output in a string: `sprintf`

The `fprintf` function is clearly the best tool available for displaying textual information on the screen, complete with handling the appropriate number of decimal places and plugging in values of different types of variables so the information can be read and interpreted by the user. In certain scenarios, it may be desired to create a formatted string, saved in MATLAB's workspace, similar to the way MATLAB can format a sentence for output to the screen. The **sprintf** function uses the same syntax as the `fprintf` function to create formatted text, complete with formatting codes and variables, but the result of the `sprintf` function is stored into a variable that will contain the formatted text as a string variable.

EXAMPLE 18-6

Assume that the following variables are defined in MATLAB's workspace:

```
Age = 20;
Average = 79.939;
Food = 'pizza';
Postal = 515;
```

To create strings of the formatted text displayed in the previous example, the `sprintf` function can be used instead of the `fprintf` function. Assume that it is desired to store the formatted strings into variables `Str1`, `Str2`, `Str3`, and `Str4`.

Since age is a decimal value, it is best to use the `%0.0f` *control character:*

```
Str1 = sprintf ('My age is %0.0f\n',Age);
```

To display the average with two decimal places:

```
Str2 = sprintf ('The average is %0.2f\n',Average);
```

To display a string within a sentence:

```
Str3 = sprintf ('My favorite food is %s\n',Food);
```

To display a postal code, the field width of a postal code is typically five numbers long:

```
Str4 = sprintf ('My postal code is %05.0f\n',Postal);
```

18.3 PLOTTING

LEARN TO: Create proper plots using `plot` and `fplot`
Create logarithmic plots with `loglog`, `semilogx`, and `semilogy`
Create multiple plots in a figure with `subplot`

MATLAB has many plotting options, not all of which concern us in this book. The simplest way to create in MATLAB is the **plot** command.

Plotting Variables

If the vectors `time` and `distance` contain paired data, `plot(time,distance)` will plot the time data on the abscissa and the distance data on the ordinate.

In addition to time and distance, assume a second distance vector, `distance2`, contains additional distance recordings at the same time values in `time`. To plot both `distance` and `distance2` on the same graph, the plot command will contain both the `distance` and `distance2` vectors: `plot(time,distance,time,distance2)`.

plot (A,B,C,...)

- ***A:*** The horizontal axis values
- ***B:*** The vertical axis values
- ***C:*** (Optional) Specification of the line/symbol type/color
- ***A,B,C:*** Can be repeated for multiple data series `(plot (A1,B1,C1,A2,B2,C2))`

Creating Proper Plots

A number of other functions allow the programmer to automatically insert information onto a graph generated in MATLAB. Many plot options are defined by separate commands, discussed below. These can be applied when the plot is created or afterwards, as long as the plot is unchanged.

- **close all:** A command to close all currently open figures. This is especially helpful when developing the code to generate your plot.
  ```
  close all
  ```
- **xlabel:** Insert a label on the abscissa.
  ```
  xlabel('Time (t) [s]')
  ```
- **ylabel:** Insert a label on the ordinate.
  ```
  ylabel('Height (H) [m]')
  ```
- **title:** Insert a title. This is optional, and should not be used if the graph will be used in another presentation format (such as a report) and a caption will accompany the graph.
  ```
  title('Flight Across The United States')
  ```
- **legend:** Insert a legend on the graph if more than one data series is present. The order of elements provided in the legend is based on the order they are passed into the plot function.
  ```
  legend('Airplane 1','Airplane 2')
  ```
- **grid:** Allows the grid to be turned on or off.
  ```
  grid on
  ```

- **axis:** Allows the setting of the minimum and maximum values on the abscissa and ordinate, in that order.

  ```
  axis([0 50000 0 90000])
  ```

  Alternately, the functions `xlim` and `ylim` can be used to define the [minimum, maximum] limits on the abscissa and ordinate, respectively.

The following examples use several special graphing commands. For more information on graphing properties, refer to the MATLAB Graphing Properties, Linespec, and Special Character reference tables contained in the endpages of your textbook.

EXAMPLE 18-7

Create a graph of multiple experimental data series for the following data:

H1 = [10, 15, 25, 35, 55]; H2 = [10, 30, 50, 70, 100];
P1 = [0.27, 0.41, 0.68, 0.95, 1.5]; P2 = [0.11, 0.33, 0.54, 0.76, 1.09];

```
clear
clc
H1=[10, 15, 25, 35, 55];    H2=[10, 30, 50, 70, 100];
P1=[0.27,0.41,0.68,0.95,1.5];
P2=[0.11,0.33,0.54,0.76,1.09];
plot(H1,P1,'o',H2,P2,'+');
axis([0 120 0 1.75]);
title('Experimental Data, Multiple Data Series Plot');
legend('Mass = 100 kg','Mass = 250 kg');
xlabel('Height (H) [m]');
ylabel('Power (P) [hp]');
grid on
```

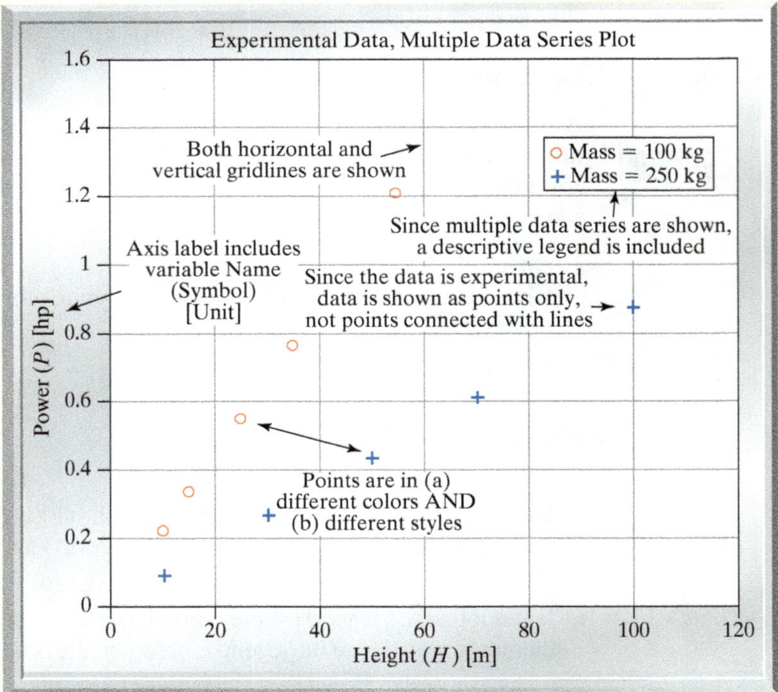

Figure 18-1 Example of a proper plot, showing multiple experimental data sets

Plotting Theoretical Expressions

Another way MATLAB can plot a function is by using **fplot**.

The way the function `fplot` works is to create a function of the variable (*x*), which takes the range [*a b*] in the following format:

```
fplot ('function(x)', [a b])
```

- ***function(x)*** is the function to be plotted, in terms of variable x.
- ***[a b]*** are the limits on the variable x.

It can only use a SINGLE VARIABLE for the function variable, such as an **x** or a **t** or an **m**. So, for example, try to run this plot:

```
fplot ('100*x + 52 / x + 11', [10 30])
```

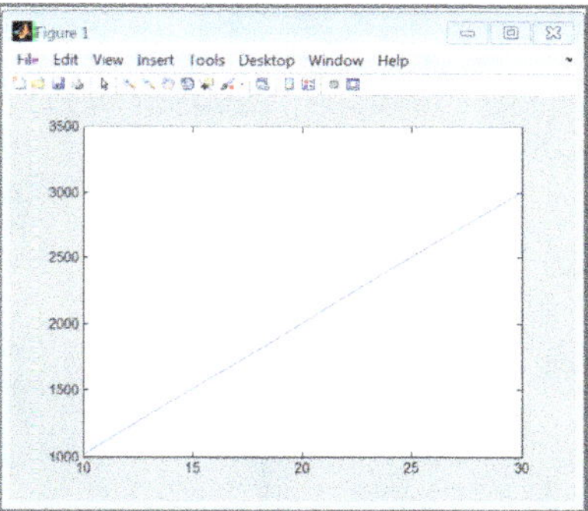

Then try this:

```
fplot ('[100*m + 52, 300*m + 100]', [10 30])
```

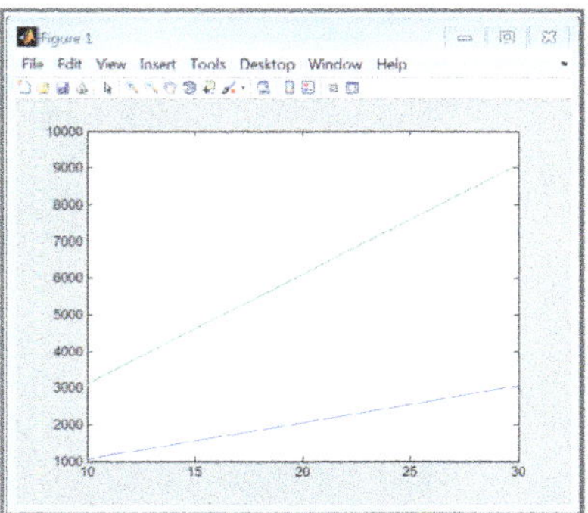

This, however, will result in an error:

```
t = 25;
x = 10;
fplot ('x*m + t', [10 30])
```

Since `fplot` is unable to use variables defined in the program, it has limited usefulness.

EXAMPLE 18-8

Create a graph of a single theoretical data series using `fplot` of the following function, from $x = 0$ to 60: $y = 0.0273\,x + 0.05$:

```
clear
clc
fplot('0.0273*x+0.05',[0 60]);
title('Theoretical Data, Single Data Series Plot');
xlabel('Mass (m) [g]');
ylabel('Density (\rho) [g/cm^3]');
grid on
```

The same graph can be created using the `plot` *function. If no marker style is specified, the program will draw a line.*

```
clear
clc
x = [0 60];
plot(x,0.0273*x+0.05);
title('Theoretical Data, Single Data Series Plot');
xlabel('Mass(m)[g]');
ylabel('Density(\rho)[g/cm^3]');
grid on
```

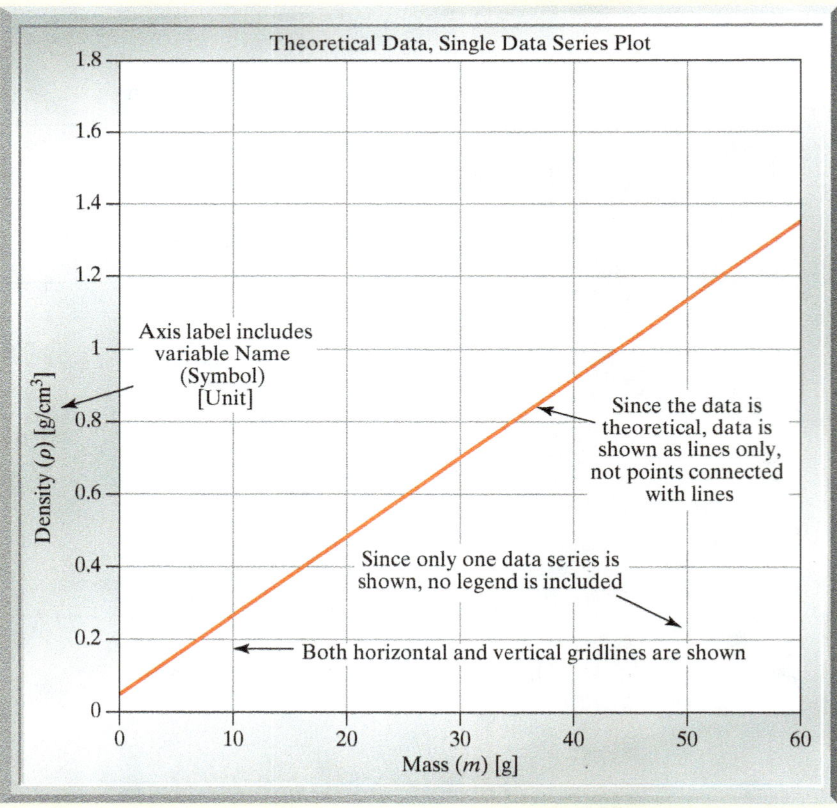

Figure 18-2 Example of a proper plot, showing a single theoretical data set.

● EXAMPLE 18-9

Create a graph of multiple experimental data series on logarithmic axes for the following data:

R1 = [10, 30, 50, 70, 100];
V1 = [0.11,0.33, 0.54, 0.76, 1.09];

R2 = [10, 15, 25, 35, 55];
V2 = [0.27, 0.41, 0.68, 0.95, 1.5];

```
clear
clc
R1=[10,30,50,70,100];
V1=[0.11,0.33,0.54,0.76,1.09];
R2=[10,15,25,35,55];
V2=[0.27,0.41,0.68,0.95,1.5];
loglog(R1,V1,'s',R2,V2,'o');
grid on
xlim([1 100]);
ylim([0.1 10]);
title('Logarithmic, Experimental, Multiple Data Series Plot');
xlabel('Radius (R) [cm]');
ylabel('Volume (V) [cm^3]');
legend('Cylinder #1','Cylinder #2')
```

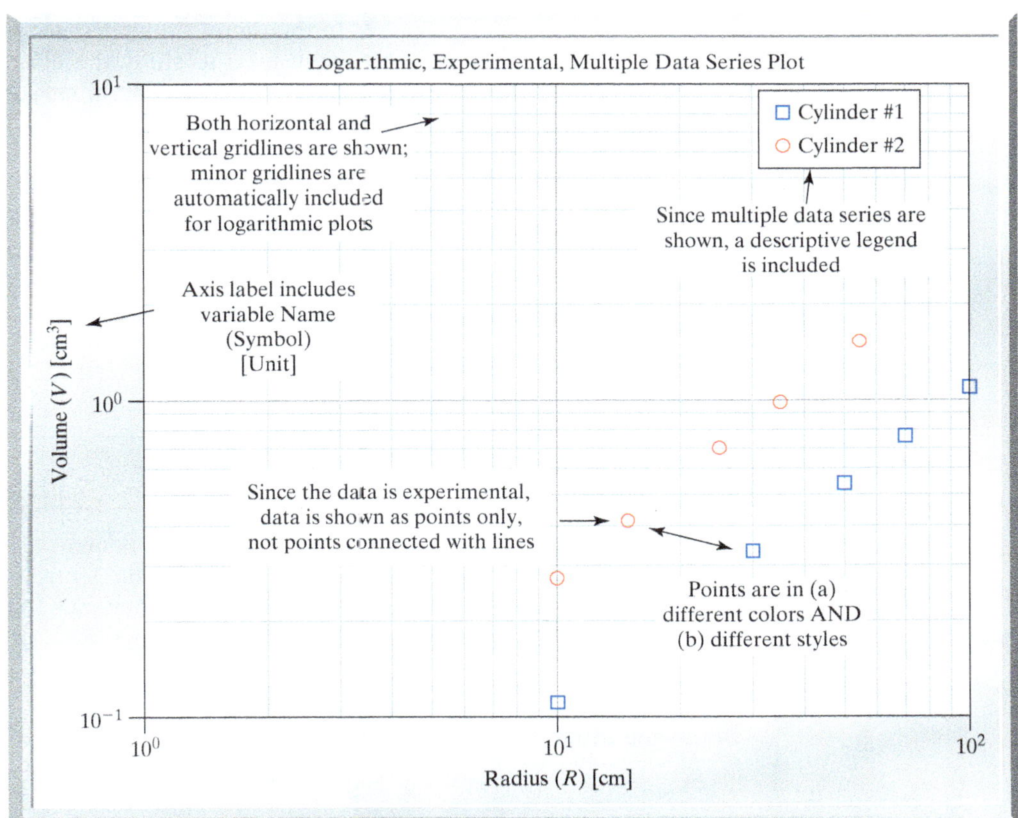

Figure 18-3 Example of a proper plot, showing multiple experimental data sets and logarithmic axes.

Creating Figures in MATLAB

> **NOTE**
> `figure` = Creates a new drawing window

When MATLAB creates a new plot, it automatically draws the plot to a "Figure" window. If a Figure window is already open, MATLAB will replace the current plot with the new plot within this window. To create multiple plot windows, type the command `figure` to create a new window before drawing your next plot. Add all titles, *x*- and *y*-axis labels, etc., before you initiate the `figure` command, because those functions execute to the newest figure. If you need to reference a previous figure, you can type `figure(N)`, where N is the number of the figure. By default, MATLAB names the first figure created Figure 1 and increases figure number by 1 as new figures are created.

It is also possible to control the background color of the figure window using the `figure` command. This may be desirable if the figure will appear in another type of publication, such as a report or presentation.

```
figure('color','white')
```

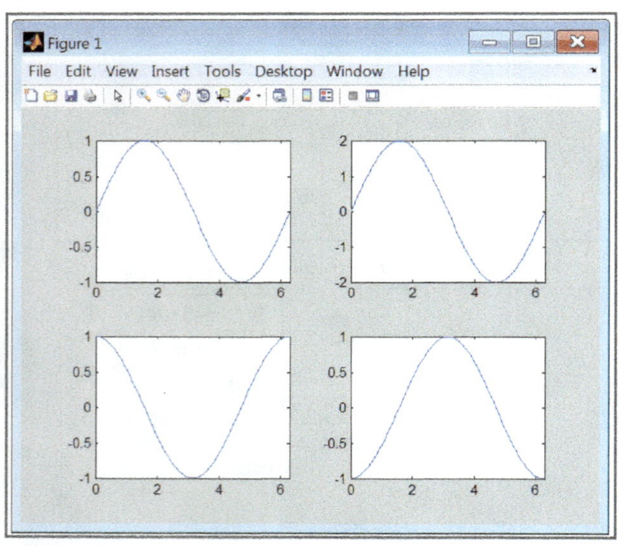

Using Subplots in MATLAB

In addition to generating single plots, MATLAB can generate multiple plots, called subplots, in a single window. This can make it easier to compare the results of different cases or to study a process from a variety of perspectives simultaneously.

In the figure shown to the left, the `fplot` command has been used to study the effect of changing the input parameters to the general sine function $A \sin(Bx + C)$. Notice that even plots in MATLAB are arranged in matrices. The command to generate the grouping of plots is shown below. Note that parameter A is amplitude, B is frequency, and C represents a phase shift.

The **subplot** command is a void function that accepts three inputs, but generates no output on the screen. As soon as the `subplot` command is executed, the next plotting command and related plotting functions will apply to that specific plot in the `subplot` until the next `subplot` command appears in the code.

subplot (R,C,N)

- ***R:*** The number of rows of plots to appear in the figure.
- ***C:*** The number of columns of plots to appear in the figure.
- ***N:*** The position in the subplot where the plot commands following the subplot command will apply.

The graph positions, N, are numbered starting in the top-left corner and going across each row, starting with the number 1 and ending with the value of R × C. The following program was used to create the subplot figure shown:

`subplot(2,2,1),fplot('sin(x)',[0 2*pi])`	plots sin(x) in top left
`subplot(2,2,2),fplot('2*sin(x)',[0 2*pi])`	plots 2 sin(x) in top right
`subplot(2,2,3),fplot('sin(x+pi/2)',[0 2*pi])`	plots sin($x + \pi/2$) in bottom left
`subplot(2,2,4),fplot('sin(x-pi/2)',[0 2*pi])`	plots sin($x - \pi/2$) in bottom right

EXAMPLE 18-10 Create a graph from the previous two examples on the same figure with 2 rows in a single column.

```
clear
clc
figure('color', 'white')
subplot(2,1,1);
fplot('0.0273*x+0.05',[0 60];
title('Theoretical Data, Single Data Series Plot');
xlabel('Mass (m) [g]');
ylabel ('Density (\rho) [g/cm^3]');
grid on
R1=[10,30,50,70,100];
V1=[0.11,0.33,0.54,0.76,1.09];
R2=[10,15,25,35,55];
V2=[0.27,0.41,0.68,0.95,1.5];
subplot(2,1,2);
loglog(R1,V1,'s',R2,V2,'o');
grid on
xlim([1 100]);
ylim([0.1 10]);
title('Logarithmic, Experimental, Multiple Data Plot');
xlabel('Radius (R) [cm]');
ylabel('Volume (V) [cm^3]');
legend('Cylinder #1', 'Cylinder #2')
```

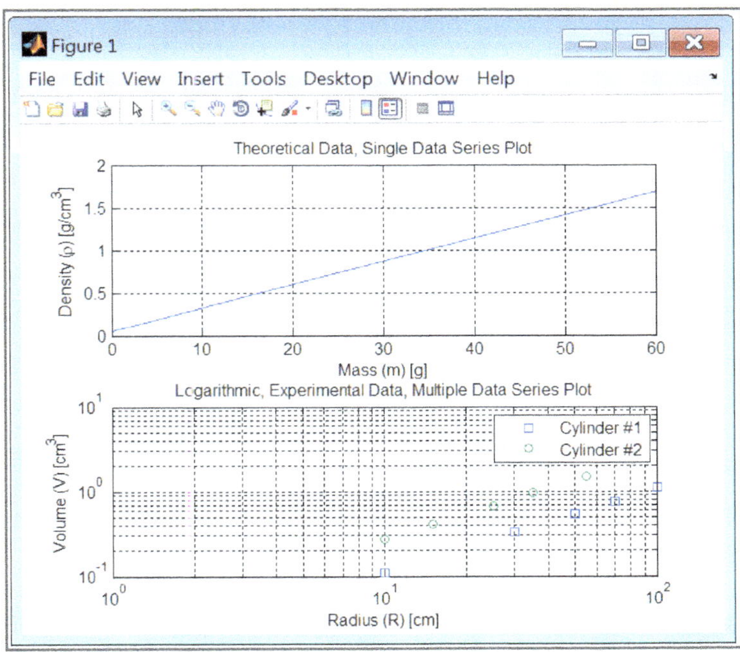

Figure 18-4 Example of a proper plot using subplots of the form 2 rows, 1 column.

18.4 POLYFIT

LEARN TO: Display trendlines with `polyfit` for linear, power or exponential trends

In the chapters on Models and Systems, three general models of importance to engineers were discussed in detail: **linear**, **power**, and **exponential**. In software programs like Microsoft Excel, built-in trendline tools will allow you to add lines representing a "best-fit" mathematical model to data sets on graphs. In MATLAB, the **polyfit** function will allow a programmer to pass in two equally sized vectors and determine the polynomial of order n that best fits the data. Consider the generic form of a polynomial:

$$p(x) = C_n X^n + C_{n-1} X^{n-1} + \cdots + C_1 X + C_0$$

In the equation above, the coefficients ($C_n, C_{n-1}, \ldots, C_1, C_0$) are returned by the `polyfit` function, given the order of the polynomial (n) in the input.

The syntax of `polyfit` is:

```
C = polyfit (X, Y, n)
```

- ***C:*** Vector of the resulting coefficients. Note that C(1), the first element of the vector C, contains the coefficient by the highest power of the variable listed in the equation, C_n.
- ***X:*** Vector of values that correspond to the abscissa
- ***Y:*** Vector (of equal length to X) of values that correspond to the ordinate
- ***n:*** is the order of the polynomial. For a linear relationship, `n = 1`.

Graphing a Trendline in MATLAB

1. Create a proper plot of the data.
2. Determine if the form of the relationship is linear, power, or exponential.
3. Use the appropriate `polyfit` command based on Step 2 to determine the parameters "m" and "b" for the trendline equation.
4. Create a theoretical data set to form the trendline. Use a reasonable range based on the experimental range for the "x" values. Use the parameters from Step 3 and the theoretical "x" values. For example: `x = [1:1:10]; y = m*x + b;`
5. Add the theoretical data set (x,y) to the graph created in Step 1.
6. Add the proper plot elements, such as the trendline equation, to complete the graph.

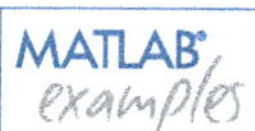

Linear Relationships

Linear relationship takes the form $y = mx + b$, where m is the slope of the line and b is the y-intercept. Linear relationships are discussed in detail in the Models and Systems chapter. The syntax of `polyfit` for a linear relationship is:

```
C = polyfit (X, Y, 1)
m = C(1)
b = C(2)
```

EXAMPLE 18-11

You are part of a firm designing nanoscale speedometers to measure speeds of small moving creatures like centipedes. To test your sensor, you place a centipede on a surface marked with a grid calibrated in millimeters and measure the following data, given in MATLAB notation. Use the data to create a graph and a mathematical model.

```
T=[0, 20, 40, 60, 80, 100, 120];       % time (t) [s]
D=[0, 105, 197, 310, 390, 502, 599];  % distance (d) [mm]
```

Step 1. *Write the code to create a graph of the experimental data set [T, D]. To save space, most proper plot commands are not included here; see Example 18-7 above.*

```
plot(T,D,'dr')
```

Step 2. *This is a linear relationship.*

Step 3. *To determine the trendline parameters using the* `polyfit` *function:*

```
C=polyfit(T,D,1)
```

Note the resulting elements in `C([4.9714, 2.1429])` *are the parameters "m" and "b" of the linear equation.*

Step 4. *Create a theoretical data set in MATLAB:*

```
T2=[0:5:120];
D2=m*T2+b;
```

Step 5. *To plot the trendline as a straight, red solid line, use the* `hold on` *function to add the trendline to the original graph:*

```
hold on
plot(T2,D2,'-r');
```

Step 6. *To add a trendline equation, use the* `sprintf` *function to convert the numbers stored in variables m and b into strings and concatenate all of the strings into a single string variable:*

```
TE=sprintf('d = %0.0f t + %0.0f',m,b);
```

To add this expression on your graph near the trendline, you will need to provide coordinates where the equation should be shown using the `text` *function. The coordinates are*

NOTE

The values for x- and y-coordinates use the actual coordinate values on the plot. The center-left end of the text string will be located at those coordinates.

determined by examining the resulting graph, and choosing a location that does not interfere with the data. Sometimes, it is easiest to create the graph, run the code, then alter the code to include the text location.

```
text(45,450,TE)
```

which take the form **`text(xcoordinate,ycoordinate,variable)`**.

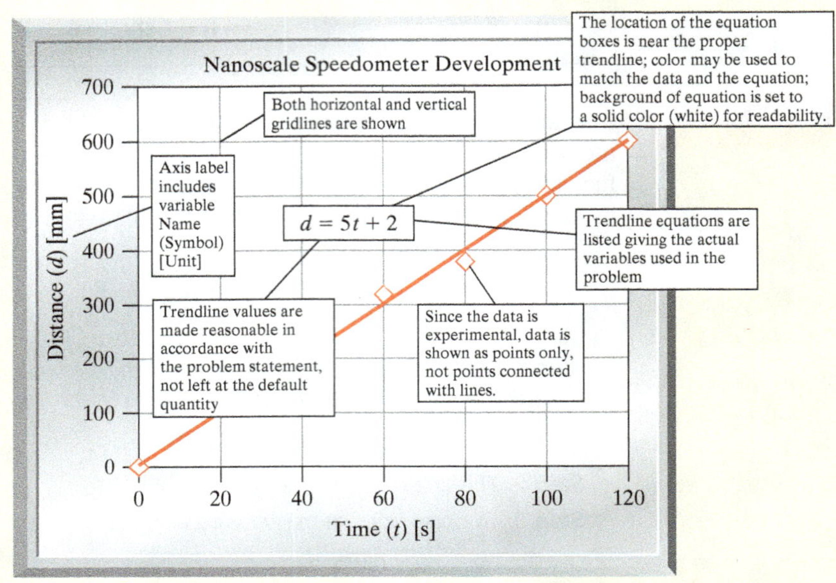

Figure 18-5 Example of a proper plot, showing single experimental data set with a linear trendline.

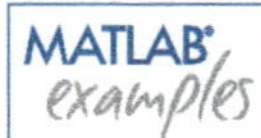

Power Relationships

A power relationship takes the form: $y = bx^m$. Unfortunately, MATLAB does not have a nice, clean function like `polyfit` for power functions, so some simple math must be performed in order to use the `polyfit` function with a power relationship. If we take the log of both sides of the general power relationship equation, we get:

$$\log(y) = m\log(x) + \log(b)$$

Power law relationships, including this derivation, are discussed in detail in the Models and Systems chapter. This expression is similar to an order-1 polynomial, so we can use `polyfit` to create a linear fit between log(x) and log(y) to calculate the value of m and log(b). Note that this calculation requires the use of the common (base 10) logarithm, not the natural logarithm. **In MATLAB, the base-10 logarithm function is `log10` and the natural logarithm function is `log`.**

NOTE

Many computer applications and calculators use `log` for the base 10 logarithm and `ln` for the natural logarithm. It is easy to forget that MATLAB is different, so be careful!

Assuming abscissa and ordinate values are stored in variables X and Y respectively:

```
C = polyfit (log10 (X), log10 (Y),1)
m = C(1)
b = 10^C(2)
```

For power relationships, it is possible to use the natural logarithm instead of the common logarithm. The difference is how you calculate the values of b.

```
C = polyfit(log(X),log(Y),1)
m = C(1)
b = exp(C(2))
```

Using either the common log or natural log for performing these calculations is correct as long as the calculation of b is handled appropriately.

EXAMPLE 18-12

Joule's first law, also known as the Joule effect, relates the heat generated to current flowing in a conductor. It is named for James Prescott Joule, the same person for whom the unit of Joule is named. The Joule effect states that the electric power (P) can be calculated as $P = I^2R$, where R is the resistance in ohms and I is the electrical current in amperes. The following data are collected in MATLAB notation. Use the data to create a graph and a mathematical model.

```
I=[0.50,1.25,1.50,2.05,2.25,3.00,3.20,3.50];     % Current (I) [A]
P=[1.2,7.5,11.25,20,25,45,50,65];                % Power (P) [W]
```

Step 1. *Write the code to create a graph of the experimental data set [I, P]. To save space, most proper plot commands are not included here;* *see Example 18-7 above.*

```
plot(I,P,'sb')
```

Step 2. *This is a power relationship.*

Step 3. *To determine the trendline parameters:*

```
C=polyfit(log10(I),log10(P),1)
m=C(1);
b=10^C(2);
```

Step 4. *Create a theoretical data set in MATLAB:*

```
I2=[0.5:0.5:3.5];
P2=b*I2^m;
```

Step 5. *To plot the trendline as a straight, blue solid line, use the* `hold on` *function:*

```
hold on
plot(I2,P2,'-b');
```

Step 6. *To add a trendline equation to your graph, use the* `sprintf` *function to convert the numbers stored in variables n and b into a single string variable:*

```
TE = sprintf('P = %3.1f I^%1.0f',b,m);
text(1,55,TE)
```

NOTE

When placing a text string on a figure, the caret (^) says to subscript the next single character or subscript the contents of braces { } immediately following.

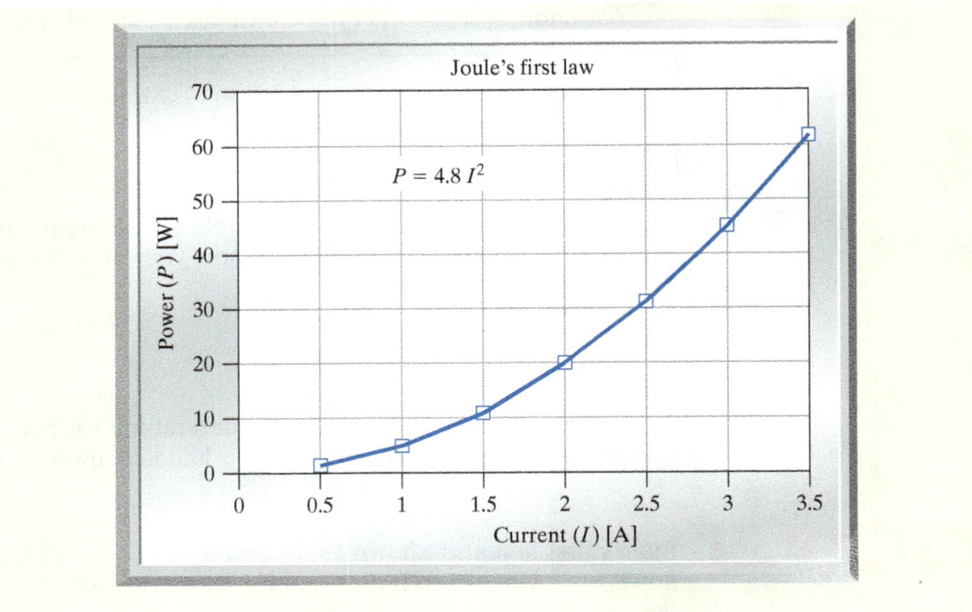

MATLAB examples

Exponential Relationships

An exponential relationship takes the form: $y = be^{mx}$. Once again, MATLAB does not have a nice, clean function like `polyfit` for exponential functions, so some math must be performed again in order to use the `polyfit` function with an exponential relationship. If we take the natural log of both sides of the general exponential relationship equation, we get:

$$\ln(y) = mx + \ln(b)$$

Experimental relationships, including this derivation, are discussed in detail in the Models and Systems chapter. This expression is similar to an order-1 polynomial, so we can use `polyfit` to create a linear fit between x and $\ln(y)$ to calculate the value of m and $\ln(b)$. Recall that the natural logarithm function in MATLAB is `log`. Assuming our abscissa and ordinate values are stored in variables X and Y, respectively:

```
C = polyfit (X, log(Y), 1)
m = C(1)
b = exp(C(2))
```

EXAMPLE 18-13

A reaction is carried out in a closed vessel. The following data are taken for the concentration of the organic material as a function of time from the start of the reaction. A proposed mechanism to predict the concentration at any given time is $C = C_0 e^{-kt}$, where k is the reaction rate constant and C_0 is the initial concentration of the species at time zero. The following data are collected in MATLAB notation. Use the data to create a graph and a mathematical model.

```
T=[36,65,100,160];                  % Time (t) [min]
C=[0.145,0.120,0.100,0.080];        % Concentration (C) [g / L]
```

Step 1. *Write the code to create a graph of the experimental data set [T, C]. To save space, most proper plot commands are not included here; see Example 18-7 above.*

```
plot(T,C,'*g')
```

Step 2. *This is an exponential relationship.*

Step 3. *To determine the trendline parameters:*

```
P=polyfit(T,log(C),1)
m=P(1);
b=exp(P(2));
```

Step 4. *Create a theoretical data set in MATLAB:*

```
T2=[20:10:160];
C2=b*exp(m*T2);
```

Step 5. *Plot the trendline*

```
hold on
plot(T2,C2,'-.g');
```

Step 6. *Add a trendline equation to the graph.*

```
TE=sprintf('C = %3.2f e^{%4.3ft}',b,m);
text(43,0.085,TE);
```

NOTE

When placing a text string on a figure, the caret (^) says to subscript the next single character or subscript the contents of braces { } immediately following.

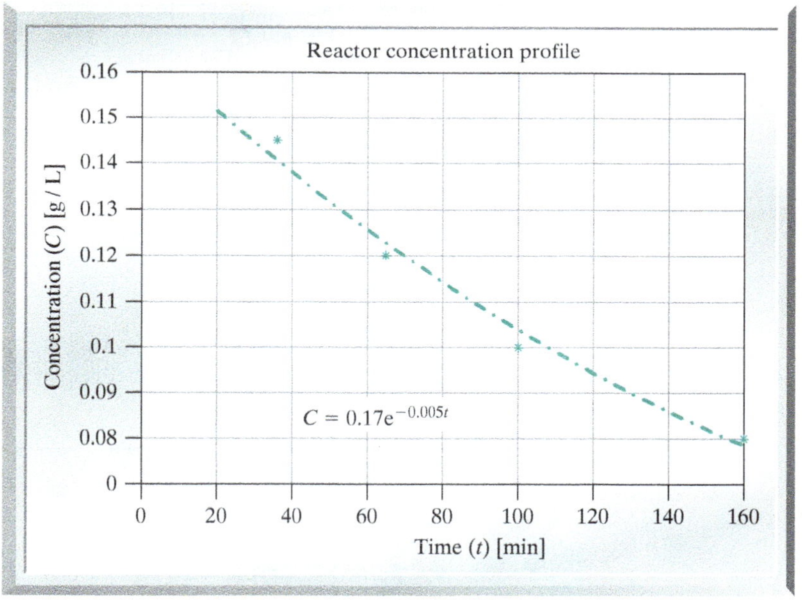

18.5 MICROSOFT EXCEL I/O

LEARN TO: Read basic Microsoft Excel workbook info with `xlsfinfo`
Read Microsoft Excel workbooks with `xlsread`
Write to Microsoft Excel workbooks with `xlswrite`

WARNING FOR MAC USERS!

Due to incompatibilities between Microsoft Excel and the Mac OS version of MATLAB 2012 and earlier, you must run at least MATLAB version 2012 in Microsoft Windows in order to use the built-in Excel I/O (input/output) functions discussed in this document. MATLAB 2013 handles Excel I/O on both platforms. In any case, Microsoft Office must be installed on the matching Operating System to use the Excel I/O functions.

In this section, you will learn to read data from Microsoft Excel workbooks into MATLAB and write information from MATLAB to an Excel workbook. File input and output is ubiquitous in computer programming if you want to give your programs and functions the ability to save computed information for use at a later time. We discuss file input and output using Microsoft Excel workbooks because it is a common file format used in many different disciplines.

Reading Microsoft Excel Workbooks

For importing data from Excel files, there are two built-in functions in MATLAB that will allow us to read in all of the information we need from our Excel files; they are `xlsfinfo` and `xlsread`.

xlsfinfo

The **xlsfinfo** function in MATLAB allows us to extract information about Microsoft Excel files stored in MATLAB's current directory. The `xlsfinfo` function will return three outputs:

```
[FileType, Sheets, ExcelFormat] = xlsfinfo (filename)
```

where

- ***FileType*** is a variable containing a string that describes the type of the file. If we call `xlsfinfo` on a file that is not a Microsoft Excel workbook, this string will be empty (`''`); otherwise, 'Microsoft Excel Spreadsheet' will be saved in the `FileType` variable. This is not particularly helpful information, but does give us a mechanism for determining whether or not the file we are trying to open is a valid Microsoft Excel workbook.
- ***Sheets*** is a cell array variable containing the names of each worksheet within the Excel workbook. Each sheet name is stored as text within the cell array.
- ***ExcelFormat*** is a variable containing a string that describes the version of the Microsoft Excel workbook. In general, this information is not useful, but if the string is blank, it is an indicator that Microsoft Excel is not installed on your computer.

- ***fileName*** is the name of the Microsoft Excel file in which we are interested. The file name must be passed into `xlsfinfo` as a string and must contain the file extension of the Microsoft Excel workbook as part of the file name. The file extension of Microsoft Excel workbooks is usually .xls or .xlsx.

The following examples will use the starter file titled "ClemsonWeather.xlsx" found in the online materials. If you open the file, you will notice that there are three worksheets within the workbook, each containing day-by-day weather information for Clemson, SC, in 2008 to 2010.

For each day, we have the following values recorded the high-, average-, and low-temperature readings measured in degrees Fahrenheit (`TempHighF`, `TempAvgF`, `TempLowF`).

xlsread

The **xlsread** function allows us to read in data from Microsoft Excel workbooks.

There are a number of different variations on how `xlsread` can be used, but we will only learn about a few of the variants involving different combinations of arguments and returned values.

If we want to tell `xlsread` to import values from a particular worksheet, we can use the following syntax:

```
D = xlsread (filename, Sheet)
```

where

- ***fileName*** is the name of the Excel workbook including the file extension, as a string.
- ***Sheet*** is a string containing the name of the Excel worksheet we want to import into our variable `D`.

EXAMPLE 18-14

Use `xlsread` to import all of the weather data from 2008 from ClemsonWeather.xlsx.

```
WD08 = xlsread('ClemsonWeather.xlsx','2008');
```

If we open `WD08` *in MATLAB's Variable Editor, we can verify that our matrix contains all of the numeric values in the "2008" worksheet, beginning at cell* `B2`*. Notice that the dimension of* `WD08` *is* 366×15*, which is what we expected since 2008 was a leap year.*

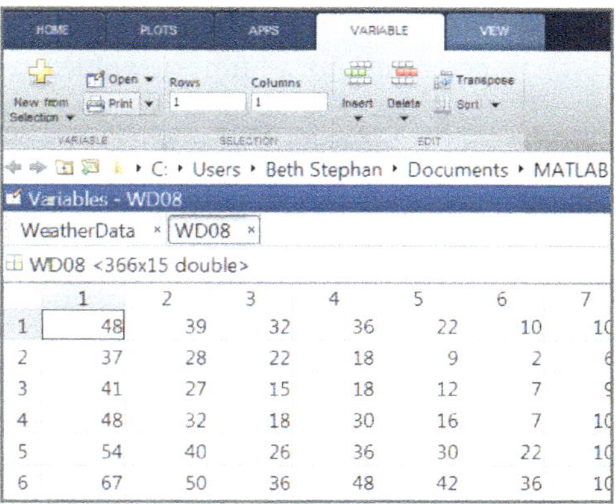

If we only want to import a particular range of an Excel file, we can add in another argument into `xlsread` that will allow us to type in ranges of values. For example, if we only wanted to import all of the cells between B2 and B5, we could provide a range B2:B5 just like we would type into Microsoft Excel.

```
D = xlsread (filename, Sheet, Range)
```

where

- ***fileName*** is the name of the Excel workbook including the file extension, as a string.
- ***Sheet*** is a string containing the name of the Excel worksheet we want to import into our variable `D`.
- ***Range*** is a string containing the cell range written in Microsoft Excel notation.

EXAMPLE 18-15

Use `xlsread` to import the first 10 high temperatures from 2009 from file ClemsonWeather.xlsx. If we open ClemsonWeather.xlsx in Microsoft Excel, we notice that the first 10 values occur between B2 and B11.

```
WD09 = xlsread('ClemsonWeather.xlsx','2009','B2:B11');
```

MATLAB creates a 10×1 matrix *containing a column of values representing the first 10 high temperatures recorded in 2009.*

If we need to extract any of the text information from the Excel Workbook, we need to modify our syntax to capture two outputs from `xlsread`. To extract nonnumeric values from an Excel Workbook:

```
[D, T] = xlsread (filename, Sheet)
```

where

- ***fileName*** is the name of the Excel workbook including the file extension, as a string.
- ***Sheet*** is a string containing the name of the Excel worksheet we want to import.
- ***D*** is a matrix of all of the numeric values in Sheet.
- ***T*** is a cell matrix of all nonnumeric (text, dates, etc.) values in the worksheet. If a cell contains a numeric value, that value will be stored as an empty string (`''`) in `T`. In other words, matrix `D` and `T` will not necessarily have the same dimensions. For more information, refer to the examples below.

It is worth noting that even though we are using the syntax requiring a sheet name to save text values, any of the previously covered notations will allow you to capture text values in an Excel worksheet.

EXAMPLE 18-16

Use `xlsread` to import the first 10 high temperatures and the corresponding text values for the dates from 2009 from ClemsonWeather.xlsx. In the worksheet, the first 10 values occur between B2 and B11 and the dates are listed between A2 and A11, so the range we need to import is A2:B11.

```
[WD09,WT09] = xlsread('ClemsonWeather.xlsx','2009','A2:B11');
```

MATLAB creates two 10 × 1 matrices *containing a column of values representing the first 10 high temperatures recorded in 2009 in* WD09 *and the corresponding text dates in a cell matrix* WT09. *In this special case, since the entire second column in* WT09 *was blank (all values were numeric), MATLAB automatically removed the second column so that* WT09 *is a* 10 × 1 matrix.

EXAMPLE 18-17

Use `xlsread` to import the numeric and nonnumeric values in the 2009 sheet in ClemsonWeather.xlsx.

```
[D09,T09] = xlsread('ClemsonWeather.xlsx','2009');
```

If we look at the dimensions of `D09` *and* `T09`, *we note that* `D09` *is a* 365 × 15 matrix, *but* `T09` *is a* 366 × 16 cell matrix. *This is due to the fact that* `T09` *contains the header row with all of the column labels, as well as the entire first column containing the corresponding dates. If we examine* `T09` *in the Variable Editor, we see that the only values in the cell matrix are the first row and first column and all of the other values in the cell matrix are empty strings (''). In light of these phenomena, special care must be taken when attempting to write MATLAB code that attempts to relate cell matrices and numeric data matrices imported using* `xlsread`.

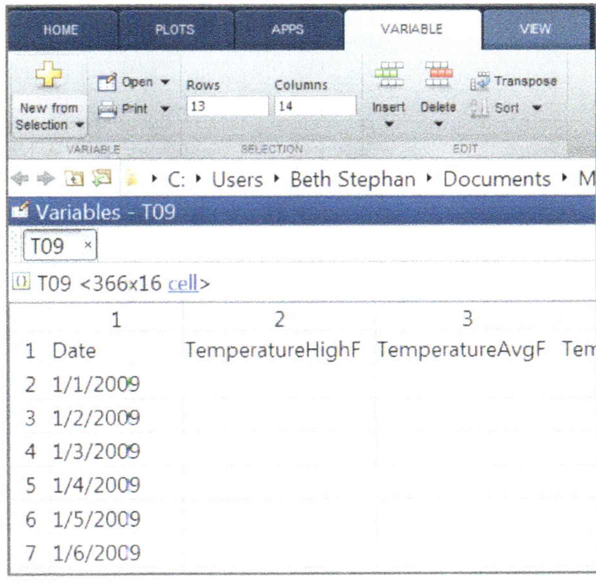

Writing Microsoft Excel Workbooks

MATLAB provides a built-in function, **xlswrite**, that enables MATLAB to export data to writing Microsoft Excel workbooks.

xlswrite

Just like with `xlsread`, `xlswrite` accepts different syntax for different writing scenarios. For the sake of brevity, we will only cover the standard use of `xlswrite`, but refer to MATLAB's help and doc documentation for more information. The function `Xlswrite` returns no output and only requires input arguments:

```
xlswrite (fileName, Matrix, Sheet, Cell)
```

where

- ***fileName*** is a string containing the file name you want to write to. If the Excel file does not exist, MATLAB will create a blank workbook with three blank worksheets ('Sheet1,' 'Sheet2,' and 'Sheet3') and display a warning "Warning: Added specified worksheet." This is common and is not considered an error message.
- ***Matrix*** is either a numeric or cell matrix containing the data you want to write to Excel file.
- ***Sheet*** is a string that contains the sheet name where you want to write your data.
- ***Cell*** is a string that designates the top-left corner of where MATLAB should start writing the data in `Matrix` in the Excel workbook. If `Cell` is not specified, MATLAB will default to writing with the top-corner set to cell A1.

EXAMPLE 18-18

Use `xlsread` to import the numeric and nonnumeric values in the 2009 sheet in ClemsonWeather.xlsx, calculate all of the average high temperatures in units of degrees Celsius, and write the output to a new Excel file called newWeather.xlsx.

```
[D09,T09] = xlsread('ClemsonWeather.xlsx','2009');
T09{1,2} = 'TempHighC';
D09(:,1) = (D09(:,1) - 32)/1.8;
xlswrite('newWeather.xlsx',T09,'2009');
xlswrite('newWeather.xlsx',D09,'2009','B2');
```

Note that we are able to use the `xlswrite` *function twice: first to write the headers and dates to the Excel file, followed by writing the matrix containing the new temperatures.*

In-Class Activities

ICA 18-1

Write a MATLAB statement that results in the input request shown in bold. The >> shows where your statement is typed, and the | shows where the cursor waits for input. The display must be correctly positioned. Each has a space before the cursor (shown as |). The input variable name and the variable type are shown at the right.

(a) `>>`
`**Enter the length of the bolt in inches:**` | (`bolt`, a number)

(b) `>>`
`**Enter the company's name:**` | (`Company`, text)

(c) `>>`

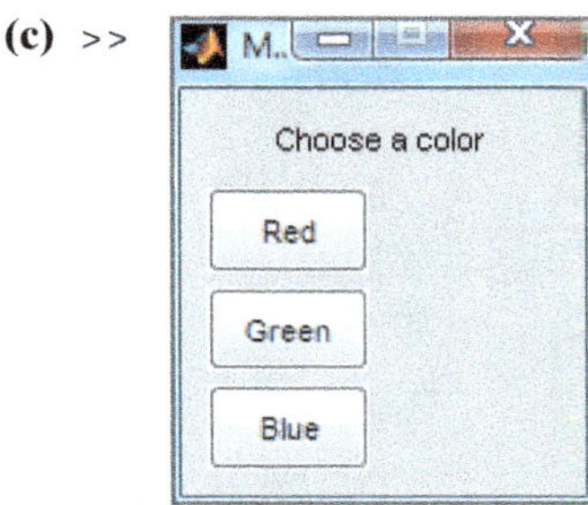

(In this statement, the window displayed will appear for the user to choose a color, and the result will be stored in the variable `LineColor`.)

ICA 18-2

You are writing code that is part of a purchasing system. For each item below, write a MATLAB statement to accomplish the task.

(a) Generate a menu that asks the user what they wish to buy and gives them a choice of three items: Flange, Bracket, or Hinge. The user's choice should be placed in a variable named **`PartType`**.

(b) Ask the user how many parts they wish to buy. Store their response in a variable named **`NumParts`**.

(c) Ask the user to enter their name. Store their response in a variable named **`BuyerName`**.

ICA 18-3

For the following questions, $z = 100/810$. Write the MATLAB output that would result from each statement.

(a) `>> disp(z)`

(b) `>> fprintf('%f',z)`

(c) `>> fprintf('%e',z)`

(d) `>> fprintf('%0.4f',z)`

ICA 18-4

For the following questions, $z = 100/810$. Write the MATLAB output statement that displays z in the format shown in bold. DO NOT hard code the values, use the value stored in z to print each one. In each case, the >> by the letter in parentheses shows where your statement is typed.

(a) `>>`
`0.1`

(b) `>>`
`0.123`

(c) `>>`
`1.235e-001`

(d) `>>`
`The value of z is 0.123.`

(e) `>>`
`z is 0.123, so 10 z is 1.235.`

ICA 18-5

Assume the following variables are stored in MATLAB's workspace:

```
MCost = 450;                 % machine cost in dollars/day
WRate = 40;                  % widgets produced/day
OCost = 1150;                % operating cost in dollars/day
WPrice = 47.87;              % sales price in dollars/widget
Days = 71;                   % number of days the machine operates
WName = 'Sonic Pliers';      % name of the widget produced
```

Determine the output displayed on the screen by the following code:

```
TCost=MCost+OCost;
fprintf('Total Cost per Day:$%0.2f\n',TCost)
NumW=WRate*Days;
fprintf('A total of %0.0f %s was produced in %0.0f days.\n',NumW,...
...WName,Days)
Income=NumW*WPrice;
fprintf('These will sell at $%0.2f each for a total of $%0.0f.\n', WPrice,...
...Income)
fprintf('This will make a profit of $%0.2f.\n',Income-TCost*Days)
```

ICA 18-6

Assume that a three-element row vector V already exists. Write a single MATLAB statement that will print the contents of V diagonally from top left to bottom right. Display each value with three decimal places.

Example: V = [42, −17.9626, 0.03654]
Sample Output:
42.000
        −17.963
                0.037

ICA 18-7

The tiles on the space shuttle are constructed to withstand a temperature of 1,950 kelvin. Write a MATLAB program that will display the temperature in the four temperature units [kelvins, degrees Celsius, degrees Fahrenheit, and degrees Rankine]. Each value should be incorporated into a sentence with appropriate text, and each sentence should appear on a new line. Format all values as integers.

ICA 18-8

The specific gravity of acetic acid (vinegar) is 1.049. Write a MATLAB program to display the density of acetic acid in units of pounds-mass per cubic foot, grams per cubic centimeter, kilograms per cubic meter, and slugs per liter. Incorporate each value into a sentence with appropriate text, each sentence on a new line. All numeric values should be given to two decimal places.

ICA 18-9

Write a MATLAB program that will allow a user to type the specific heat of a value in calories per gram degree Celsius. Display the converted value in units of British thermal units per pound-mass degree Fahrenheit in the Command Window the following format using one decimal place: "The specific heat is ____ BTU / (lb_m*deg F)."

ICA 18-10

Write a program that will allow the user to type a liquid evaporation rate in units of kilograms per minute and display the value in units of pounds-mass per second, slugs per hour, and grams per second. Display the result in the Command Window using one decimal place: "The evaporation rate is _____ pounds-mass per second, ____ slugs per hour, or ____ grams per second."

ICA 18-11

In order to calculate the pressure in a flask, write a program that allows the user to type the volume of the flask in liters, the amount of an ideal gas in units of moles, and the temperature of the gas in kelvins. Display the result in the Command Window in the following format, using one decimal place: "The pressure is ___ atmospheres."

ICA 18-12

Write a program that will allow the user to type the input power of the motor in watts, the mass of the object in kilograms, the height the object will be raised in meters, and the time it took to raise the object in seconds. Display the calculated value in the Command Window in the following format, using one decimal place: "The motor is ____% efficient."

ICA 18-13

Write a MATLAB program that will ask the user to enter their age, then ask them to enter the name of their best friend, and then ask them to enter their friend's age. Determine the difference in the ages, and express this value as a positive integer. Finally, a statement in the following format should be displayed in the command window:

```
My age is ____ years.
__________ is my best friend. My friend's age is _____ years.
The difference is our age is _____ years.
```

ICA 18-14

Write a program that asks a user to enter, one at a time, the four numbers to fill a 2 × 2 matrix. For each number entered, make sure the user knows the location in the matrix of the number being entered. When the function begins, the following information should display for the user:

```
Function asks the user for a 2 X 2 matrix and display the result.
Please enter only integers, between -99,999 and 99,999.
```

Display the matrix in the Command Window as two rows and two columns, with the integers in the columns evenly spaced apart, using field width to control the column spacing.

ICA 18-15

Joule's first law, also known as the **Joule effect**, relates the heat generated to current flowing in a conductor. It is named for James Prescott Joule, the same person for whom the unit of joule is named. Create a proper plot of the experimental data.

Current (*I*) [A]	0.50	1.25	1.50	2.25	3.00	3.20	3.50
Power (*P*) [W]	1.20	7.50	11.25	25.00	45.00	50.00	65.00

ICA 18-16

Create a proper plot of the following set of experimental data collected during the charging of a capacitor. In this plot, time should be on the abscissa.

Time (*t*) [s]	0.2	0.4	0.6	0.8	1.0
Voltage (*V*) [V]	75.9	103.8	114.0	117.8	119.2

ICA 18-17

The lumen [lm] is the SI unit of luminous flux, a measure of the perceived power of light. To test the power usage, you run an experiment and measure the following data. Create a proper plot of these data, with electrical consumption (EC) on the ordinate.

	Electrical Consumption [W]	
Luminous Flux [lm]	Incandescent 120 Volt	Compact Fluorescent
80	16	
200		4
400	38	8
600	55	
750	68	13
1,250		18
1,400	105	19

ICA 18-18

You want to create a graph showing the theoretical relationship of an ideal gas between pressure (P) and temperature (T). Assume the tank has a volume of 12 liters and is filled with nitrogen (formula, N_2; molecular weight, 28 grams per mole). Allow the initial temperature to be 270 kelvin at a pressure of 2.5 atmospheres. Create a proper plot, showing the temperature on the abscissa from 270 to 350 kelvin.

ICA 18-19

The decay of a radioactive isotope can be theoretically modeled with the following equation, where C_0 is the initial amount of the element at time zero and k is the decay rate of the isotope. Create a proper plot of the decay of Isotope A [k = 1.48 hours]. Allow time to vary on the abscissa from 0 to 5 hours with an initial concentration of 10 grams of Isotope A.

$$C = C_0 e^{-t/k}$$

ICA 18-20

Create a proper plot of the theoretical voltage decay of a resistor-capacitor circuit:

$$V(t) = V_0 e^{-\frac{t}{RC}}$$

You may assume that you have a capacitance (C) of 500 microfarads [μF], a resistance (R) of 0.5 ohms [Ω], and an initial voltage (V_0) of 10 volts [V]. The plot should start at time 0 seconds and increase by intervals of 1 second to 600 seconds.

ICA 18-21

Plot the following functions as assigned by your instructor using subplots, choosing an appropriate layout for the number of functions displayed. The independent variable (angle) should vary from 0 to 360 degrees.

(a) sin(u)
(b) 3 sin(u)
(c) sin(3u)
(d) sin(u) − 3
(e) sin(u + 90)
(f) 3 sin(2u) − 2

ICA 18-22

Plot the following functions as assigned by your instructor using subplots, choosing an appropriate layout for the number of functions displayed. The independent variable (angle) should vary from 0 to 360 degrees.

(a) cos(u)
(b) −2 cos(u)
(c) cos(2u)
(d) cos(u) + 2
(e) cos(u − 45)
(f) 3 cos(2u) − 2

ICA 18-23

If an object is heated, the temperature of the body will increase. The energy (Q) associated with a change in temperature (ΔT) is a function of the mass of the object (m) and the specific heat (C_p). In an experiment, heat is applied to the end of an object, and the temperature change at the other end of the object is recorded. This leads to the equation shown below.

$$\Delta T = \frac{Q}{mC_p}$$

An unknown material with a mass of 5 kilograms is tested in the lab, yielding the following results. Use the `polyfit` function to determine the specific heat of this material and store the final result in the variable C_p.

Create a proper plot of the data. Add a linear trendline, showing the resulting trendline equation, on the graph for a change in temperature over a range of energy from 5 joules to 70 joules.

Heat Energy Applied (Q) [J]	17	40	58
Temperature Change (ΔT) [K]	2	5	7

ICA 18-24

The resistance of a typical carbon film resistor will decrease by about 0.05% of its stated value for each degree Celsius increase in temperature. Silicon is very sensitive to temperature, decreasing its resistance by about 7% for each degree Celsius increase in temperature. This can be a serious problem in modern electronics and computers since silicon is the primary material from which many electronic devices are fabricated.

Create a proper plot to compare a carbon film resistor with a resistor fabricated from specially doped silicon ("doped" means impurities such as phosphorus or boron have been added to the silicon).

For relatively small temperature differences from the reference temperature, this process is essentially linear. Use `polyfit` to determine linear models for each data set. For each model, add the trendline and the associated trendline equation to the graph. Use an appropriate location for the equations to clearly associate them with the correct trendline.

Temperature (*T*)[°C]	Resistance (*R*) [Ω]	
	Carbon Film	Doped Silicon
15	10.050	10.15
20	10.048	9.85
25	10.045	9.48

ICA 18-25

Today, most traffic lights have a delayed green, meaning there is a short time delay between one light turning red and the light on the cross-street turning green. An industrial engineer has noticed that more people seem to run red lights that use delayed green. She conducts a study to determine the effect of delayed green on driver behavior. The following data were collected at several test intersections with different green delay times. These data represent only those drivers who continue through the intersection when the light turns red *before* they reach the limit line, defined as the line behind which a driver is supposed to stop. The data show the "violation time," defined as the average time between the light turning red and the vehicle crossing the limit line, as a function of how long the delayed green has been installed at that intersection.

Create a proper plot of the violation time (V, on the ordinate) and the time after installation (t, on the abscissa) for all three intersections on a single graph.

Use `polyfit` to determine linear models for each data set. For each model, add the trendline and the associated trendline equation to the graph. Use an appropriate location for the equations to clearly associate them with the correct trendline.

Time After Installation (*t*) [months]	Violation Time (*V*) [s]		
	Intersection 1 1-Second Delay	Intersection 2 2-Second Delay	Intersection 3 4-Second Delay
2	0.05	0.1	0.5
5	0.1	0.5	1.5
8	0.3	1	2.5
11	0.4	1.3	3.1

ICA 18-26

Cadmium sulfide (CdS) is a semiconducting material with a pronounced sensitivity to light—as more light strikes it, its resistance goes down. In real devices, the resistance of a given device may vary over four orders of magnitude or more. An experiment was set up with a single light source in an otherwise dark room. The resistance of three different CdS photoresistors was measured when they were at various distances from the light source. The farther they were from the source, the dimmer the illumination on the photoresistor.

Create a proper plot of the data. Use `polyfit` to determine power model for each data set. For each model, add the trendline and the associated trendline equation to the graph. Use an appropriate location for the equations to clearly associate them with the correct trendline.

Distance from Light (d) [m]	Resistance (R) [Ω]		
	A	B	C
1	79	150	460
3	400	840	2,500
6	1,100	2,500	6,900
10	2,500	4,900	15,000

ICA 18-27

Your supervisor has assigned you the task of designing a set of measuring spoons with a "futuristic" shape. After considerable effort, you have come up with two geometric shapes that you believe are really interesting.

You make prototypes of five spoons for each shape with different depths and measure the volume each will hold. The table below shows the data you collected.

Depth (d) [cm]	Volume of Shape	
	A (VA) [mL]	B (VB) [mL]
0.5	1	1.2
0.9	2.5	3.3
1.3	4	6.4
1.4	5	7.7
1.7	7	11

Create a proper plot of the data. Use `polyfit` to determine power models for each data set. For each model, add the trendline and the associated trendline equation to the graph. Use an appropriate location for the equations to clearly associate them with the correct trendline.

Use your models to determine the depths of a set of measuring spoons comprising the following volumes for each of the two designs. Use your models to determine the depths of a set of measuring spoons comprising the following volumes for each of the two designs. `V = [¼ tsp, ½ tsp, ¾ tsp, 1 tsp, 1 Tbsp]`. NOTE: 1 Tbsp = 3 tsp

Print the results to the Command Window in a table similar to the format shown below.

```
Volume Needed (V) [tsp]        0.25    0.5    0.75    1    3
Depth of Design A (dA) [cm]
Depth of Design B (dB) [cm]
```

ICA 18-28

Three different diodes were tested: a constant voltage (0.65 volts) was held across each diode while the current through each was measured at various temperatures. The following data were obtained.

Create a proper plot of the data. Use `polyfit` to determine exponential models for each data set. For each model, add the trendline and the associated trendline equation to the graph.

Temperature (T) [K]	Current (I_D) [mA]		
	Diode A	Diode B	Diode C
275	852	2,086	264
281	523	1,506	179
294	194	779	81
309	69	390	35
315	47	301	26

ICA 18-29

If a hot liquid in a container is left to cool, its temperature (T) [°C] will gradually approach room temperature. This model will have the form $T = A + B\,e^{mt}$. In this case, m will always be negative and its dimension will be inverse time. If it is not clear how to determine A and B, consider that $T = A + B$ when $t = 0$, and $T = A$ as t approaches infinity.

You are to write a FUNCTION named `Cooling` that will accept two parameters:

- `Tr`: Room temperature in °C
- `TData`: A two row matrix containing measurements of a cooling liquid at various times. The first row contains the times of the measurements in minutes, and the second row contains the corresponding temperature measurements in °C

The function must perform the following operations:

- Plot the measured points
- Determine an exponential model to fit the data
- Add the exponential trendline to the graph
- Add the trendline equation for the exponential model to the graph

Example:

Room Temperature: `Tr = 30`

Temperature Data: `TData = [0.5 3 7 11; 85 48 35 33]`

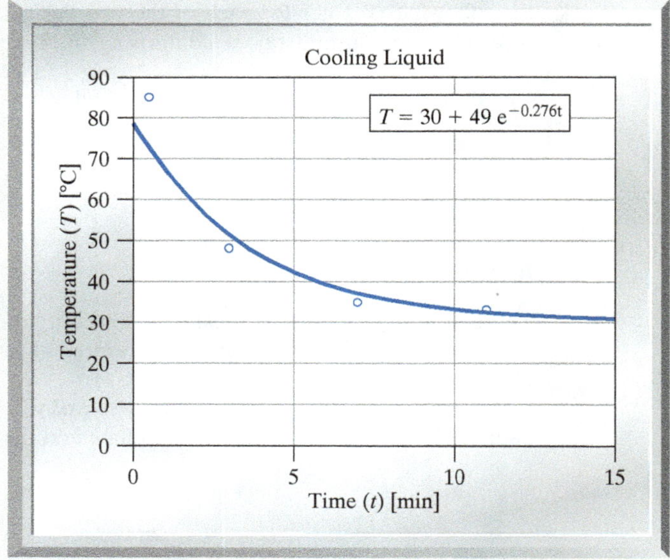

ICA 18-30

You are an engineer working for M & M / Mars™ Corporation in the M&M plant. For Halloween, M&Ms are produced in "fun size". To help with quality control, you create the worksheet below. The factory workers will examine sample bags of M&Ms, and enter the weight of the bag and the individual count of M&Ms contained in the bag.

Online, you have been given the following data in a Microsoft Excel workbook called **CandyCount.xlsx** with the data stored on a sheet named "M and M data". Only a portion of the actual data is shown.

	A	B	C	D	E	F	G	H
1	Bag Number	Mass of Bag (m) [g]	Red	Orange	Yellow	Green	Blue	Brown
2	1	15.0	3	1	5	4	4	1
3	2	13.9	3	2	3	5	1	2
4	3	14.7	1	1	9	2	3	1
5	4	14.3	2	5	4	2	1	3
6	5	14.7	[illegible]	[illegible]	6	[illegible]	[illegible]	[illegible]

Write a MATLAB program to read in the Excel data, calculate the data required as shown in the output table below, and write the results to a new Microsoft Excel file. Your output should appear as follows, where the highlighted portion is replaced by the values you calculate in your solution. Your program should also use formatted `fprintf` statements to display the table in the command window, filling in the missing information (highlighted) and displaying the data using a reasonable number of decimal places.

The program should assume the user could could enter more data to the original worksheet or modify the color names on the worksheet and the program should still run correctly.

	A	B	C	D	E	F	G
1	M and M Analysis by Color						
2		Red	Orange	Yellow	Green	Blue	Brown
3	Average per bag:						
4	Average mass per bag:						
5	Average total per bag:						
6	Color that appears the most:						
7	Color that appears the least:						

Chapter 18 REVIEW QUESTIONS

1. The specific gravity of gold is 19.3. Write a MATLAB program that will ask the user to input the mass of a cube of solid gold in units of kilograms and display the length of a single side of the cube in units of inches. The output should display a sentence like the one shown below, with the length formatted to 2 decimal places.

Sample Input/Output:

```
Enter the mass of the cube [kilograms]: 0.4
The length of one side of the cube is 1.08 inches.
```

2. An unmanned X-43 A scramjet test vehicle has achieved a maximum speed of Mach number 9.68 in a test flight over the Pacific Ocean. Mach number is defined as the speed of an object divided by the speed of sound. Assuming the speed of sound is 343 meters per second, write a MATLAB program to determine speed in units of miles per hour. Your program should ask the user to provide the speed as Mach number and return the speed in miles per hour in a formatted sentence, displayed as an integer value, as shown in the sample output below.

Sample Input/Output:

```
Enter the speed as a Mach number: 9.68
The speed of the plane is 7425 mph.
```

3. A rod on the surface of Jupiter's moon Callisto has a volume measured in cubic meters. Write a MATLAB program that will ask the user to type in the volume in cubic meters in order to determine the weight of the rod in units of pounds-force. The specific gravity is 4.7. Gravitational acceleration on Callisto is 1.25 meters per second squared. The input and output of the program should look similar to the output below. Be sure to report the weight as an integer value.

Sample Input/Output:

```
Enter the volume of the rod [cubic meters]: 0.3
The weight of the rod is 397 pounds-force.
```

4. The Eco-Marathon is an annual competition sponsored by Shell Oil, in which participants build special vehicles to achieve the highest possible fuel efficiency. The Eco-Marathon is held around the world with events in the United Kingdom, Finland, France, Holland, Japan, and the United States.

A world record was set in the Eco-Marathon by a French team in 2003 called Microjoule with a performance of 10,705 miles per gallon. The Microjoule runs on ethanol. Write a MATLAB program to determine how far the Microjoule will travel in kilometers given a user-specified amount of ethanol, provided in units of grams. Your program should ask for the mass using an input statement and display the distance in a formatted sentence similar to the output shown below.

Sample Input/Output:

```
Enter mass of ethanol [grams]: 100
The distance the Microjoule traveled is 577 kilometers.
```

5. Write a function and program to determine the mass of oxygen gas (formula: O_2, (molecular weight = 32 grams per mole) in a container in units of grams. The function accept from the program input arguments representing the volume of the container in units of gallons, the temperature in the container in degrees Celsius, and the pressure in the container in units of atmospheres in the Command Window. The function output should be a single formatted string that contains the mass of the oxygen gas formatted to 1 decimal place followed by the units. For example the strings '10.0 grams' or '13.3 grams' would be samples of the expected output of your function. Note that this function should not generate any output on the screen.

The program should prompt the user to input the volume in gallons, the temperature in degrees Celsius, and the pressure in atmospheres, and then call the function to create the formatted string containing the mass. Finally, your program should contain an output statement to display the mass of the oxygen gas in the container. For your test case, you may assume that the user provides 1.25 gallons for the volume of the container, 125 degrees Celsius for the temperature, and 2.5 atmospheres for the pressure in the container.

Sample Input/Output:

```
Enter the volume [gallons]: 1.25
Enter the temperature [deg C]: 125
Enter the pressure [atm]: 2.5
The mass of the oxygen gas in the container is 11.6 grams.
```

6. Write a program and function to calculate a temperature provided by the user in units of Fahrenheit to units of kelvin. Your function should contain a single input argument, the temperature in degrees Fahrenheit, and return a single string as output. The string should be a formatted string containing the temperature followed by the units (K), where the value of the temperature is formatted to contain 0 decimal values. As an example, valid outputs of this function would be strings like '100 K' or '273 K'.

The program should allow the user to type in the temperature in degrees Fahrenheit, then pass that value into the aforementioned function, and display the output in a sentence formatted as shown below. As a test case, you may assume the user provides the temperature of −129 degrees Fahrenheit, which is the world's lowest recorded temperature.

Sample Input/Output:

```
Enter the temperature [deg F]: -129
The equivalent temperature is 184 K.
```

7. You are part of an engineering firm on contract by the U.S. Department of Energy's Energy Efficiency and Renewable Energy task force to develop a program to help consumers measure the efficiency of their home appliances. Your job is to write a program that measures the efficiency of stove-top burners. Before using your program, the consumer will place a pan of room temperature water on their stove (with 1 gallon of water), record the initial room temperature in units of degrees Fahrenheit, turn on the burner, and wait for it to boil. When the water begins to boil, they will record the time in units of minutes it takes for the water to boil. Finally, they will look up the power for the burner provided by the manufacturer.

The output of your program should look like the output displayed below; where the highlighted values are example responses typed by the user into your program. Note that your code should line up the energy and power calculations, as shown below. In addition, your code must display the efficiency as a percentage with one decimal place and must include a percent symbol.

```
Household Appliance Efficiency Calculator: Stove

Type the initial room temperature of the water [deg F]: 68
Type the time it takes the water to boil [min]: 21
Type the brand name and model of your stove: Krispy 32-Z
Type the power of the stove-top burner [W]: 1200

Energy required:        1267909 J
Power used by burner:   1006 W

Burner efficiency for a Krispy 32-Z stove: 83.9%
```

Sample Input/Output:

Stove Model	Room Temp [°F]	Time to Boil [min]	Rated Burner Power [W]
Krispy 32-Z	68	21	1,200
MegaCook 3000	71	25	1,300
SmolderChef 20F	72	21	1,500
Blaze 1400-T	68	26	1,400
CharBake 5	69	18	1,350

8. We want to conduct a breakeven analysis for a wooden baseball bat manufacturer who is interested in diversifying their product line. As a reminder breakeven analysis was discussed previously in the Graphing Solutions chapter, specifically in Example 11-9. Currently, the manufacturer produces white ash bats; they are interested in purchasing a second machine line to produce either maple or bamboo baseball bats.

To help the manufacturer look at the different scenarios, write a program that asks the user of the program to input the following variables IN THIS SPECIFIC ORDER:

- Selling price of a maple bat (in dollars)
- Selling price of a bamboo bat (in dollars)
- The total number of bats the manufacturer can produce per week; this value is the same for either type of bat
- The number of weeks the manufacturer plans to run their bat production machinery in a year

The manufacturing process can only make a single type of bat each week, and will produce the same number of bats in a week regardless of the material used.

Revenue is determined as the selling price of an object times the number of objects. Your program should display the total revenue generated for the scenario, as shown below:

```
Producing ## bats a week for ## weeks in a year will generate:
    Maple bat revenue: $####.##
    Bamboo bat revenue: $####.##

Total number of bats produced: #.###e+##
```

The revenue should be displayed with two decimal places and the total number of bats produced should be displayed in exponential notation with three decimal places.

In addition, each revenue line should display with a single tab at the front of each sentence, as shown above. The "#" character in the output displayed above will be actual numbers in the program you create.

Sample Input/Output. The highlighted values are entered by the user:

```
Type the selling price of a maple bat (in dollars): 16
Type the selling price of a bamboo bat (in dollars): 24
Type total number of bats manufacturer can produce per week: 50
Type number of weeks manufacturer plans to run production: 50

Producing 50 bats a week for 50 weeks will generate:
    Maple bat revenue: $40000.00
    Bamboo bat revenue: $60000.00
Total number of bats produced: 2.500e+03
```

9. We want to conduct a breakeven analysis for a wooden baseball bat manufacturer who is interested in diversifying their product line. As a reminder breakeven analysis was discussed previously in the Graphing Solutions chapter, specifically in Example 11-9. Currently, the manufacturer produces white ash bats; they are interested in purchasing a second machine line to produce hickory baseball bats.

To help the manufacturer look at the different scenarios, write a program that asks the user of the program to input the following variables IN THIS SPECIFIC ORDER:

- The number of weeks the manufacturer plans to run their bat production machinery in a year
- The total number of bats manufacturer can produce per week; value is same for either type of bat produced
- Material cost to produce a single white ash bat (in dollars)
- Material cost to produce a single hickory bat (in dollars)
- Selling price of a white ash bat (in dollars)
- Selling price of a single hickory bat (in dollars)

The cost for energy and labor for white ash bats is $2.75 per bat; however, due to difficulties associated with producing hickory bats, the cost for energy and labor is $4.25 per hickory bat. In your program, the cost of energy and labor should appear as variables.

Since the equipment already exists, there is no fixed cost for the white ash bats; the fixed cost for the hickory bats is $5,000 to purchase the new equipment. In your program, both fixed costs should appear as variables. Even though the fixed cost for the white ash is zero, you should still include it as a variable to allow the user to alter the program easily if, for instance, the manufacturer would want to examine upgrading the equipment.

The manufacturing process can only make a single type of bat each week, and will produce the same number of bats in a week regardless of the material used.

Your program should display the total profit generated for the scenario, as shown below:

```
Producing ## bats a week for ## weeks in a year will generate:
    White Ash bat profit: $####
    Hickory bat profit: $####
Total number of bats produced: #.#e+###
```

The profit should be displayed with no decimal places and the total number of bats produced should be displayed in exponential notation with one decimal places. Each profit line should display with a single tab at the front of each sentence. The "#" character in the output displayed above will be actual numbers in the program you create.

The program should generate two proper plots with the number of bats produced on the abscissa and the revenue and total cost on the ordinate for each material type. These two graphs should appear on the same figure using a subplot.

Finally, the program should generate a third plot that displays only the profit of each bat type with respect to the number of bats produced.

Sample Input / Output. The highlighted values are entered by the user:

```
Type number of weeks manufacturer plans to run production: 50
Type total number of bats manufacturer can produce per week: 50
Type a Material cost, white ash bat (in dollars): 8
Type a Material cost, hickory bat (in dollars): 12
Type selling price, white ash bat (in dollars): 16
Type selling price, hickory bat (in dollars): 24

Producing 50 bats a week for 50 weeks will in a year will generate:
White ash bat profit: $40000
Hickory bat profit: $60000

Total number of bats produced: 2.5e+03
```

Sample Figures of Cost and Revenue Curves:

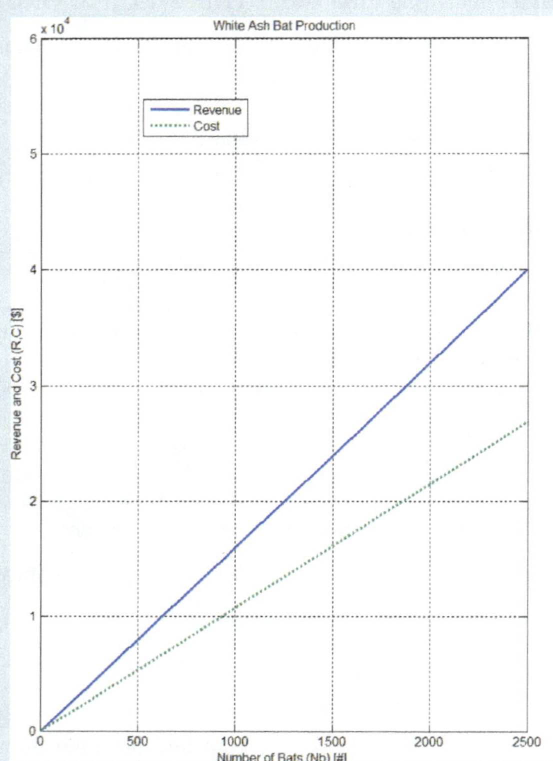

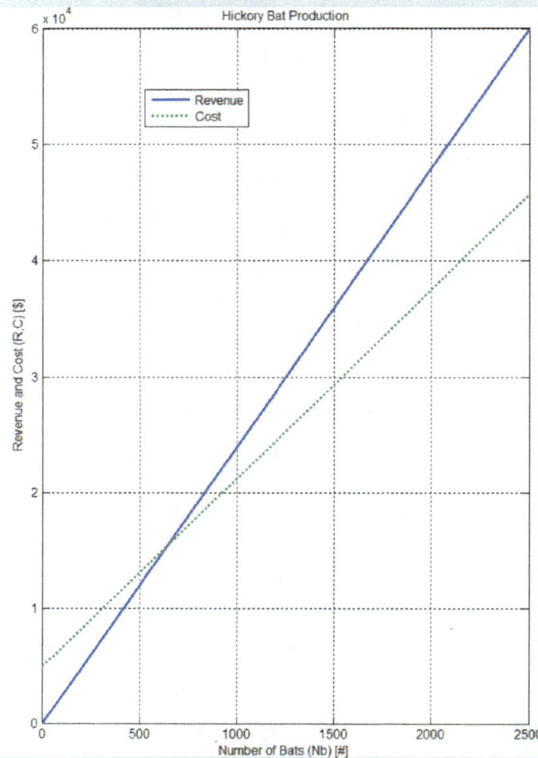

10. When one attempts to stop a car, both the reaction time of the driver and the braking time must be considered. Create a proper plot the following data.

Vehicle Speed (v) [mph]	Distance (d) [m]	
	Reaction (d_r)	Braking (d_b)
20	6	6
30	9	14
40	12	24
50	15	38
60	18	55
70	21	75

If an object is heated, the temperature of the body will increase. The energy (Q) associated with a change in temperature (ΔT) is a function of the mass of the object (m) and the specific heat (C_p). In an experiment, heat is applied to the end of an object, and the temperature change at the other end of the object is recorded. This leads to the theoretical relationship shown. An unknown material is tested in the lab, yielding the following results:

$$\Delta T = \frac{Q}{mC_p}$$

Heat applied (Q) [J]	12	17	25	40	50	58
Temp change (ΔT) [K]	1.50	2.00	3.25	5.00	6.25	7.00

11. Create a proper plot of the experimental temperature change (ΔT, ordinate) versus the heat applied (Q, abscissa).

12. Use `polyfit` to determine a linear relationship for the data set and graph the resulting trendline along with the experimental data.

13. Create a proper plot of the theoretical model that represents the temperature change (ΔT, ordinate) versus the heat applied (Q, abscissa) using `fplot`. Consider the mass (m) to be 5 kilograms and the specific heat (C_p) to be copper at 0.39 joules per gram kelvin.

Capillary action draws liquid up a narrow tube against the force of gravity as a result of surface tension. The height the liquid moves up the tube depends on the radius of the tube. The following data were collected for water in a glass tube in air at sea level.

Radius (r) [cm]	0.01	0.05	0.10	0.20	0.40	0.50
Height (H) [cm]	14.0	3.0	1.5	0.8	0.4	0.2

14. Create a proper plot of the height (H, ordinate) versus the radius (r) assuming the data are experimental.

15. Use `polyfit` to determine a power relationship for the data set and graph the resulting trendline along with the experimental data.

16. In a different experiment with different data, you determine the relationship to be $H = 0.25\ r^{-1}$. Create a proper plot of this expression using the `fplot` command, assuming the units for both the radius and height are centimeters.

17. Use `polyfit` to determine a power relationship for the data set and plot the resulting relationship on a graph with the experimental data. This plot should be a proper plot with a logarithmic axis.

In a turbine, a device used for mixing, the power required depends on the size and shape of the impeller. In the lab, we have collected the following data:

Diameter (D) [ft]	0.5	0.75	1	1.5	2	2.25	2.5	2.75
Power (P) [hp]	0.004	0.04	0.13	0.65	3	8	18	22

18. Create a proper plot of the power (P, ordinate) versus the diameter (D) assuming the data are experimental.

19. Use `polyfit` to determine a power relationship for the data set and graph the resulting trendline along with the experimental data.

20. In a different experiment with different data, you determine the relationship to be $P = 0.25\ D^5$. Create a proper plot of this expression using the `fplot` command, assuming the units for both the diameter and power are the same as shown in the table.

21. Use `polyfit` to determine a power relationship for the data set and plot the resulting relationship on a graph with the experimental data. This plot should be a proper plot with a logarithmic axis.

A pitot tube is a device that measures the velocity of a fluid, typically the airspeed of an aircraft. The failure of the pitot tube was credited as the cause of Austral Lineas Aéreas flight 2553's crash in October 1997. The pitot tube had frozen, causing the instrument to give a false reading of slowing speed. As a result, the pilots thought the plane was slowing down, so they increased the speed and attempted to maintain their altitude by lowering the wing slats. Actually, they were flying at such a high speed that one of the slats ripped off, causing the plane to nosedive; the plane crashed at a speed of 745 miles per hour [mph].

In the pitot tube, as the fluid moves, the velocity creates a pressure difference between the ends of a small tube. The tubes are calibrated to relate the pressure measured to a specific velocity, using the speed as function of the pressure difference (P, in units of pascals) and the density of the fluid (ρ, in units of kilograms per cubic meter).

$$v = \left(\frac{2}{\rho}\right)^{0.5} P^{m}$$

22. Create a proper plot of the velocity (v, ordinate) versus the pressure (P) assuming the data are experimental.

23. Use `polyfit` to determine the power relationships for the data sets and graph the resulting trendlines along with the experimental data.

24. Use `polyfit` to determine a power relationship for the data set and plot the resulting relationship on a graph with the experimental data. This plot should be a proper plot with a logarithmic axis.

Pressure (P) [Pa]	50,000	101,325	202,650	250,000	304,000	350,000	405,000	505,000
Velocity fluid A (v) [m/s]	11.25	16.00	23.00	25.00	28.00	30.00	32.00	35.75
Velocity fluid B (v) [m/s]	9.00	12.50	18.00	20.00	22.00	24.00	25.00	28.00
Velocity fluid C (v) [m/s]	7.50	11.00	15.50	17.00	19.00	20.00	22.00	24.50

A growing field of inquiry that both holds great promise and poses great risk for humans is nanotechnology, the construction of extremely small machines. Over the past couple of decades, the size that a working gear can be made has consistently gotten smaller. The table shows milestones along this path.

Years from 1967	0	5	7	16	25	31	37
Minimum gear size [mm]	0.8	0.4	0.2	0.09	0.007	2E-04	8E-06

25. Create a proper plot of the gear size (ordinate) versus the number of years from 1967 assuming the data are experimental.

26. Use `polyfit` to determine the exponential relationship for the data set and graph the resulting trendline along with the experimental data.

27. In a different experiment with different data, you determine the relationship to be MGS $= 2.5\,e^{-0.5t}$. Create a proper plot of this expression using the `fplot` command, assuming the units for both the time and gear size are the same between the two data sets.

28. Use `polyfit` to determine an exponential relationship for the data set and plot the resulting relationship on a graph with the experimental data. This plot should be a proper plot using a logarithmic axis.

29. Download the weekly retail gasoline and diesel prices Excel workbook associated with this review question and place it in your main MATLAB Current Directory. This data set is based on the data set available from the U.S. Energy Information Administration: http://www.eia.gov/dnav/pet/pet_pri_gnd_dcus_nus_w.htm.

(a) Calculate the average, minimum, and maximum retail fuel prices for each of the different types of fuel (regular, midgrade, premium, diesel) over the duration of the entire sample set. Your output should work in such a way that if the original Excel file were modified to include more weeks (rows), you would not need to change your MATLAB code. Your output should appear similar to the format below, where the blanks are replaced by the actual calculated values:

Average Weekly Retail Gasoline and Diesel Prices

	Regular	Midgrade	Premium	Diesel
Min:				
Max:				
Average:				

Your code should calculate the values shown in the output—you should not hard code the values in the output. Each value you display should appear with two decimal values. You may use any built-in MATLAB function, including functions that find the minimum, maximum, or average values.

(b) We have decided that we want to modify our previous analysis in part **(a)** to export the computed max, min, and average values to a new Microsoft Excel workbook. The data itself should be exported to a sheet named "Fuel Price Analysis." Your data should appear similar to the worksheet below. Much like part **(a)**, your code should be written in such a way that if the original Excel file were modified to include more weeks (rows), you would not need to change your MATLAB code.

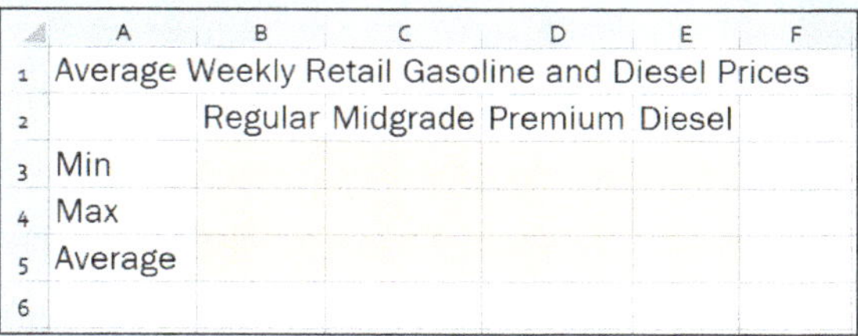

	A	B	C	D	E	F
1	Average Weekly Retail Gasoline and Diesel Prices					
2		Regular	Midgrade	Premium	Diesel	
3	Min					
4	Max					
5	Average					
6						

30. A sample of the data provided in the Microsoft Excel file online is shown below. The file contains energy consumption data by energy source per year in the United States, measured in petaBTUs.

Year	Fossil Fuels	Elec. Net Imports	Nuclear	Renewable
2007	101.605	0.106	8.415	6.830
2006	99.861	0.063	8.214	6.922

(a) Write the MATLAB code necessary to read the Microsoft Excel file and store each column of data into different variables. Create the following:

- `Yr`: A vector of all of the years in the worksheet.
- `FF`: A vector of all of the fossil fuels for each year in Yr.
- `ENI`: A vector of all electric imports for each year in Yr.
- `Nuc`: A vector of all nuclear energy consumption for each year in Yr.
- `Ren`: A vector of all renewable energy consumption for each year in Yr.
- `Hdr`: A cell array of all of the headers in row 1.

You may not hard code these variables—they should be imported from the Excel file.

(b) Create a new variable, **`TotalConsumption`**, which contains the sum of the four columns of energy consumption data for each year. In other words, since all five of the vectors (**`Yr, FF, ENI, Nuc,`** and **`Ren`**) we created in part **(a)** have the same length, the new variable, **`TotalConsumption`**, should have the same length. You may assume that you have correctly defined the variables in part **(a)**.

(c) Calculate the average fossil fuel consumption in the entire data set. For this code, you may assume that you have correctly defined **`FF`**, the variable containing the fossil fuel consumption data, in part **(a)**. Display the result of the calculation in the format shown below, where the number is shown to two decimal places:

```
The average fossil fuel consumption is ____________
petaBTU.
```

31. An Excel file named Dart Tosses.xlsx, saved in a worksheet named "Darts" contains the measurements in this data set contain the horizontal and vertical distance from the bullseye of a dart board from 20 different tosses of a dart. A portion of the data is shown below. Write the MATLAB code necessary to determine the darts that were the closest and the furthest from the bullseye. The program should tell the user in the command window, using formatted output, the darts that are closest and farthest from the bullseye.

Dart	X	Y
Dart 1	4.04	0.55
Dart 2	2.63	0.35
Dart 3	1.10	2.97
Dart 4	4.89	5.60

CHAPTER 19
LOGIC AND CONDITIONALS

Outside the realm of computing, logic exists as a driving force for decision making. Logic transforms a list of arguments into outcomes based on a decision. Some examples of everyday decision making are as follows:

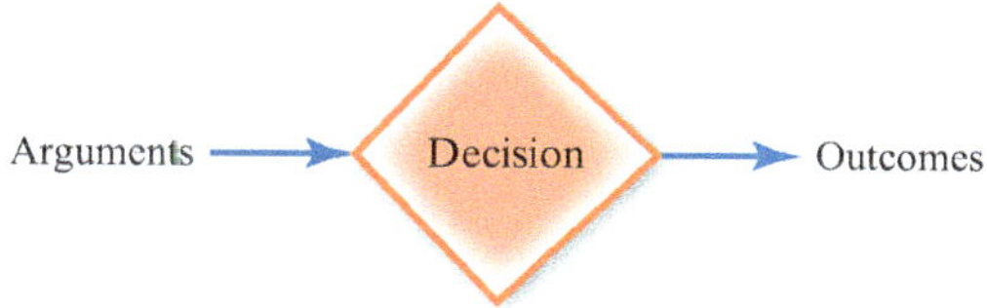

- If the traffic light is red, stop. If the traffic light is yellow, slow down. If the traffic light is green, go.

  Argument: three traffic bulbs

  Decision: is bulb lit?

  Outcomes: stop, go, slow

- If the milk has passed the expiration date, throw it out; otherwise, keep the milk.

  Argument: expiration date

  Decision: before or after?

  Outcomes: garbage, keep

To bring decision making into our perspective on problem solving, we need to first understand how computers make decisions. **Boolean logic** exists to assist in the decision-making process, where each argument has a binary result and the overall outcome exhibits binary behavior. **Binary behavior**, depending on the application, is any sort of behavior that results in two possible outcomes. Boolean logic, Boolean algebra, and related terms are named for George Boole (1815–1864), the English mathematician whose writings formed the basis of modern computer science.

19.1 RELATIONAL AND LOGICAL OPERATORS

LEARN TO: Create decisions using relational operators
Create decisions connected by logic operators
Explain the difference between short-circuit and element-wise operators

Operator	Meaning
>	Greater than
<	Less than
>=	Greater than or equal to
<=	Less than or equal to
==	Equal to
~=	Not equal to

In order to determine the relationship between two values (numbers, variables, etc.), we have a few operators with which we can compare two variables to determine whether or not the comparison is true or false.

These **relational operators** are usually placed between two different variables or mathematical expressions to determine the relationship between the two values. This expression of variable–operator–variable is typically called a **relational expression**, the result of which is always either true or false. We will refer to such results – values that can only be true or false – as **Boolean arguments**.

Sometimes all we care about is a simple relationship between two things, for example, is A greater than B? In other cases, we need to ask more complicated questions; for example is A both greater than B and less than C? Simple relational expressions can be combined by **logical operators** to create a **logical expression**. If no logical operator is required in a particular decision, then the single relational expression can be the logical expression.

To connect all of the Boolean arguments to make a logical decision, we turn to a few operators with which we can relate our arguments to determine a final outcome.

NOTE
And = &&
Or = ||
Not = ~

- **And:** The AND logical operator connects two Boolean arguments and returns the result as true if and only if *both* Boolean arguments have the value of TRUE. In MATLAB, we use the two ampersands (&&) symbol (SHIFT+7 on a keyboard) to represent the AND logical operator.
- **Or:** The OR logical operator connects two Boolean arguments and returns the result as true if *either one* (or both) of the Boolean arguments has the value of TRUE. In MATLAB, we use two pipes (||) symbol (SHIFT+\ on a keyboard) to represent the OR logical operator.
- **Not:** The NOT logical operator inverts the value of a single Boolean argument or the result of another Boolean operation. In MATLAB, we use the tilde (~) symbol (SHIFT+ ' on a keyboard) to represent the NOT logical operator. In other words, NOT True equals False, and vice versa.

EXAMPLE 19-1

Express the mathematical inequality in MATLAB code.

Mathematical Inequality	Expression
$4 \leq X < 5$	4 < = X && X < 5
$10 < X \leq 20$	10 < X && X < = 20
$30 \leq X \leq 100$	30 < = X && X < = 100

In this example and others, MATLAB cannot operate with multiple relational operators in a single expression—a common situation in mathematical inequalities. Instead, we must separate out the two relations and combine the expressions with the "and" symbol.

COMPREHENSION CHECK 19-1

What are the relational expressions for the following mathematical inequalities?
(a) $5 \leq t < 10$
(b) $-30 < M \leq 20$
(c) $Y \neq 100$

&&, &, and + ||, |, or: Say What?

By design, MATLAB contains a couple of different operators and functions for performing the "and" and "or" operations. The following set of guidelines will help you choose which operator will be the most appropriate choice for your code. In all of the following, assume the variable `x` contains the string `'tigers'`.

AND

- **&&:** The double-ampersand operator is often referred to in MATLAB documentation as a short-circuit AND operator because it will terminate the evaluation of a logical expression as soon as it encounters the first "false" in a logical expression. This is particularly useful when checking to see if a certain condition is false before continuing with a special computation that is part of the logical expression.

Sample: `isnumeric(x) && x + 10 > 5`

In the sample above, the overall result will be false because the first computation (`isnumeric(x)`) was false, so the remainder of the expression was not evaluated.

- **&:** Single-ampersand operator is often referred to in MATLAB documentation as an element-wise AND operator because it will evaluate every element in a logical expression.

Sample: `isnumeric (x) & x + 10 > 5`

In the sample above, the overall result will be an array of six false values. The reason the result is an array is because the first computation (`isnumeric(x)`) was false, so the expression will always be false, because FALSE & ANYTHING = FALSE. In this particular case, the value of x is a string, so the logical expression is computing the ASCII value of the letters t, i, g, e, r, and s, adding 10, checking to see if the result is greater than 5, and then using the AND result, which will be false, all six times.

- **`and()`:** The and function is a function that behaves similarly to the AND function in Microsoft Excel. This function behaves in the same fashion as the single-ampersand element-wise operator.

Sample: `and(isnumeric (x), x + 10 > 5)`

In the sample above, the overall result will be an array of six false values since the AND function behaves identically to the element-wise operator.

OR

- **||:** The double-pipe operator is often referred to in MATLAB documentation as a short-circuit OR operator because it will terminate the evaluation of a logical expression as soon as it encounters the first "true" in a logical expression. This is particularly useful when checking to see if a certain condition is true before continuing with a special computation.

Sample: `~isnumeric(x)||x + 5 > 0`

In the sample above, the overall result will be true because the first computation `~isnumeric(x)` was true, so the remainder of the expression was not evaluated.

- **|:** The single-ampersand operator is often referred to in MATLAB documentation as an element-wise `OR` operator because it will evaluate every element in a logical expression.

  Sample: `~isnumeric(x)|x + 5 > 0`

  In the sample above, the overall result will be an array of six true values. The reason the result is an array is because the first computation `(~isnumeric(x))` was true, and since the remainder of the expression `(x + 5 > 0)` will also be true for all 6 characters in the string `'tigers'`, the result is an array of 6 true outcomes.

- **`or` ():** The `or` function is a function that behaves similarly to the OR function in Microsoft Excel. This function behaves in the same fashion as the single-ampersand element-wise operator.

  Sample: `or (~isnumeric (x), x + 5 >0)`

  In the sample above, the overall result will be an array of six true values since the `or` function behaves identically to the element-wise operator.

The `&&` and || operators only work if the logical values being combined are logical SCALARS. If you wish an element-wise AND or OR of two matrices, you must use either `&` or | or use the `and/or` functions.

19.2 LOGICAL VARIABLES

LEARN TO: **Create and use logic variables in decisions**
Define a logic matrix
Design logic within the constraints of operator priority

Logical Scalar Variables

When two individual items are compared using relational operators, the result is either a 1, representing true; or 0, representing false. In many cases, the results of such operations are used immediately by a MATLAB command, and are not actually stored in a variable.

The results of such comparisons can, however, be assigned to a variable and used later. Although the `if` statement will not be introduced until the next section, the following example should be simple enough to understand as an example.

EXAMPLE 19-2

Assume that `A` and `B` are scalars previously defined in the workspace. We wish to write a segment of MATLAB code to determine if variable `A` is not equal to variable `B`.

```
% Option 1
if A ~=B
    fprintf('A and B are not equal')
end
```

The expression `A ~= B` is evaluated, giving a 1 if true or a 0 if false. This result is not stored in a variable, but is immediately used by the `if` statement to decide whether or not to print the message. If we wanted to know what the result of the comparison was later, we would *have to ask the question again, although* `A` *or* `B` *might have changed in the interim so we might get a different result.*

```
% Option 2
AB=A~=B;
if AB
    fprintf('A and B are not equal')
end
```

The result of the relational question is stored in the logical variable `AB`*, and will have a value of 0 or 1. The icon shown beside* `AB` *in the workspace will be a square with a checkmark, indicating it is a logical variable.*

The program will determine if the variable `AB` *contains a 1 (true), and then the message will be printed; if* `AB` *contains a 0 (false), the message will not be printed. Note that we still have access to the result of the question later since we stored the result in* `AB`*. We could, for example, use this in conjunction with asking the same question later to see if the status of the relationship between* `A` *and* `B` *had changed.*

Logical Matrices

Relational comparisons can be made between a matrix and a scalar. In this case, the result is a logical matrix with the same dimensions as the matrix being compared. Each element of this logical matrix contains the result (1 or 0) of the comparison between the scalar and the corresponding element of the matrix.

EXAMPLE 19-3

Determine the result of the following code, assuming `C` $= \begin{bmatrix} 1 & 0 & -1 \\ -2 & 2 & 3 \end{bmatrix}$

```
C0 = C <= 0;
```

The code will examine each element in the matrix, and compare the value with zero. If the value is less than or equal to zero, the number 1 will be stored in the new `C0` *matrix, in the same location as the element being compared. If the value is greater than zero, the number 0 will be stored. For example, if we compare* `C(1,1)` *– a value of 1 – to zero, this value is greater than zero so the number 0 is stored in* `C0(1,1)`.

The resulting matrix will be: `C0` $= \begin{bmatrix} 0 & 1 & 1 \\ 1 & 0 & 0 \end{bmatrix}$

Relational comparisons can also be made between two matrices, as long as they have the same dimensions. In this case, the result is a logical matrix with the same dimensions as the two matrices being compared. Each element of this logical matrix contains the result (1 or 0) of the element-wise comparisons between corresponding elements of the two matrices.

EXAMPLE 19-4

Determine the result of the following code, assuming

$$C = \begin{bmatrix} 1 & 0 & -1 \\ -2 & 2 & 3 \end{bmatrix} \text{ and } D = \begin{bmatrix} 2 & -2 & -1 \\ -1 & 2 & 1 \end{bmatrix}$$

```
CD = C <= D;
```

The code will examine each element in the matrix `C`*, and compare the value to the same element position in matrix* `D`*. If the value in* `C` *is less than or equal the value in* `D`*, the number 1 will be stored in the new* `CD` *matrix, in the same location as the elements being compared. If the value in* `C` *is greater than the value in* `D`*, the number 0 will be stored.*

For example, if we compare the element `C(1,1)` *to* `D(1,1)`*, the value of 1 is less than the value of 2 so the number 1 is stored in* `CD(1,1)`.

The resulting matrix will be: $CD = \begin{bmatrix} 1 & 0 & 1 \\ 1 & 1 & 0 \end{bmatrix}$

Relational comparisons can be used with text strings as long as the strings have the same number of characters. Note that case matters: the lowercase letter "a" is not equal to the uppercase letter "A." Since text strings are essentially vectors containing text instead of numbers, the result of such a comparison is a logical vector containing the same number of elements as each of the two strings being compared.

Stings are compared based on the ASCII code used to represent them, so you need to know the codes assigned to the various characters, especially punctuation, to understand the results of greater than or less than comparisons. The table below shows each ASCII character and the corresponding decimal value for each letter or number. Note that there are many more characters available in the full set of ASCII values (http://www.asciitable.com/). A comprehensive list can be found in the end-pages of this text.

Character	A	B	C	D	E	F	G	H	I	J	K	L	M
Value	65	66	67	68	69	70	71	72	73	74	75	76	77

Character	N	O	P	Q	R	S	T	U	V	W	X	Y	Z
Value	78	79	80	81	82	83	84	85	86	87	88	89	90

Character	a	b	c	d	e	f	g	h	i	j	k	l	m
Value	97	98	99	100	101	102	103	104	105	106	107	108	109

Character	n	o	p	q	r	s	t	u	v	w	x	y	z
Value	110	111	112	113	114	115	116	117	118	119	120	121	122

Character	0	1	2	3	4	5	6	7	8	9
Value	48	49	50	51	52	53	54	55	56	57

You may have noticed that when variables are alphabetized in the workspace, those that begin with a capital letter come before those that begin with a lowercase letter.

If a comparison between two text strings involves greater than or less than (as opposed to just equal or not equal) the results may be surprising. For example `'a' < 'A'` is false, very likely not what you expected. On the other hand, alphabetic characters of the same case do compare as you probably expect: a is "less than" b and Y is "less than" Z. Also, non-alphanumeric characters can be problematic with greater than or less than.

EXAMPLE 19-5

Determine the result of the following code, assuming

```
T = 'AbCd1;  2 # 3';  S = 'abCD3: 2, 1';
TS = T <= S;
```

The code will examine each element in the vector `T`*, and compare the value to the same element position in vector* `S` *If the ASCII code value in* `T` *is less than or equal the ASCII code value in* `S`*, the number 1 will be stored in the new* `TS` *matrix, in the same location as the elements being compared. If the ASCII code value in* `T` *is greater than the ASCII code value in* `S`*, the number 0 will be stored.*

For example, if we compare the element `T(1)` *to* `S(1)`*, the text "A" has an ASCII code value of 65 and the text "a" has an ASCII code value of 97. Since 65 is less than 97, the number 1 is stored in* `TS(1)`*.*

The resulting matrix will be: `TS = [1 1 1 0 1 1 0 1 1 0]`

Calculations using logical variables

To determine if strings of different lengths are the same, you can use `strcmp` or `strcmpi`. In this case, the result is a logical scalar—true or false.

Logical matrices allow you to do some moderately complicated operations very easily in MATLAB. Since logical matrices contain actual numeric values (although only zeros and ones), they can be used in computations with regular numeric matrices.

EXAMPLE 19-6

Assume $Q = \begin{bmatrix} 1 & -7 & 9 & -12 \\ -24 & 10 & 6 & 100 \end{bmatrix}$. Write a code segment to modify `Q` by dividing by 2 every element with a magnitude greater than or equal to 10, and leaving the other elements unchanged.

We can approach this problem by creating two matrices: one containing a value of 1 for all the elements with a magnitude greater than or equal to 10 [`BigQ`*], and one containing a value of 1 for all the elements less than 10 [*`SmallQ`*]. Then, we can recombine these two matrices for our end result.*

```
BigQ=abs (Q) >= 10;              % Find elements with magnitude ≥ 10
BigQAdjusted =0.5*BigQ.*Q;       % Divide big elements by 2
SmallQ =~BigQ.*Q;                % Remove big elements from Q
Q = BigQAdjusted+SmallQ;         % Combine the two matrices
```

The result: $Q = \begin{bmatrix} 1 & -7 & 9 & -6 \\ -12 & 5 & 6 & 50 \end{bmatrix}$

Note that this will work with any size matrix: for a 1000 × 2000 matrix, the code would be the same.

If an arithmetic operation, such as dividing by 2 or adding 6, is performed on a logical matrix, the result is a numeric matrix.

Priority of operators

The priority of operators must be observed carefully when mixing arithmetic, relational, and logical operators in the same expression. Particularly note that the arithmetic operators have priority over the relational operators, and the relational operators have priority over the logical operators. The one exception is **logical negation** (~ or `not (A)`), which has a priority equal to that of unary minus. A unary minus (−x) replaces the variable with the additive inverse; for example if x = 3, then −x = −3. The order of operator priorities is summarized in Table 19-1. Remember that parentheses have priority over everything.

Table 19-1 Priority of mixed operations

Priority	Operations
1	Transpose, Power
2	Unary Minus, Logical Negation
3	Multiplication, Division
4	Addition, Subtraction
5	Colon operator
6	Relational operators `(<, < =, >, >=, ~=, ==)`
7	`&`
8	`|`
9	`&&`
10	`||`

Functions associated with logical variables

Many of the functions used with numeric matrices can also be used with logical matrices. The table below contains a few additional functions that you might find useful. For each function, examples are given using the following variables:

```
A = [1  3  5  7];  C = [2  4  5  7];  R = A~= B;      T = B == C;
B = [2  4  6  8];  D = [2  3  5  8];  S = A + 1~= B;  U = C == D;

    Canine1='Dog';                 Canine2='Wolf';
```

Table 19-2 Selected functions used with logical variables

Function	Description
`all`	Are all elements in a given dimension of a matrix 1 (or non-zero)? Examples: `H = all (R);` `% H contains 1` `I = all (T);` `% I contains 0`
`any`	Are any elements in a given dimension of a matrix 1 (or non-zero)? Examples: `J = any (S);` `% J contains 0` `K = any (T);` `% K contains 1`
`isequal`	Determine if a group of matrices are all identical Examples: `F = isequal (B,C);` `% F contains 0` `G = isequal (A,B-1);` `% G contains 1`
`islogical`	Determine if a variable is a logical variable Examples: `D = islogical (A);` `% D contains 0` `E = islogical (R);` `% E contains 1`

`strcmp`	Case sensitive string comparison; Examples:
	`CType1 = strcmp (Canine1,Canine2);` `% CType1 contains 0`
	`CType2 = strcmp (Canine1,'Dog');` `% CType2 contains 1`
	`CType3 = strcmp (Canine1,'dog');` `% CType3 contains 0`

`strcmpi`	Case insensitive string comparison; Examples:
	`CType4 = strcmpi (Canine1,Canine2);` `% CType4 contains 0`
	`CType5 = strcmpi (Canine1,'Dog');` `% CType5 contains 1`
	`CType6 = strcmpi (Canine1,'dog');` `% CType6 contains 1`

`xor`	Exclusive OR: True if there is one 1 and one zero in corresponding elements Examples: `L = xor (T,U);` `% L contains [0 1 1 0]`

COMPREHENSION CHECK 19-2

$$\texttt{M1} = \begin{bmatrix} 6 & -3 \\ -7 & 0 \\ 2 & 10 \end{bmatrix} \quad \text{and} \quad \texttt{M2} = \begin{bmatrix} 9 & -5 \\ -4 & 1 \\ 2 & 0 \end{bmatrix}$$

(a) What is placed in `R1` by the statement `R1 = M1 <= M2`?

(b) What is placed in `R2` by the statement `R2 = (M2 < M1).*M2`?

(c) What is placed in `R3` by the statement `R3 = M2 < M1.*M2`?

COMPREHENSION CHECK 19-3

Assume `M1` and `M2` are defined as given in CC 19-2 above. Write a short section of MATLAB code that will create a matrix `R4` in which each element is the larger of the corresponding elements in `M1` and `M2`.

For the values listed above, the result should be: `R4` $= \begin{bmatrix} 9 & -3 \\ -4 & 1 \\ 2 & 10 \end{bmatrix}$

19.3 CONDITIONAL STATEMENTS IN MATLAB

LEARN TO: Write an algorithm incorporating decision making in MATLAB code
Predict the number of possible outcomes given variability in a condition
Determine if two conditional statements are logically equivalent

Conditional statements are commands by which users give decision-making ability to the computer. Specifically, the user asks the computer a question framed in conditional statements, and the computer selects a path forward (it provides an answer—either numerical or as a text comment) based on the answer to the question. Sample questions are given below:

- If the water velocity is fast enough, switch to an equation for turbulent flow!
- If the temperature is high enough, reduce the allowable stress on this steel beam!
- If the pressure reading rises above the red line, issue a warning!
- If your grade is high enough on the test, state: You Passed!

In these examples, the comma indicates the separation of the condition and the action that is to be taken if the condition is true. The exclamation point marks the end of the statement. Just as in language, more complex conditional statements can be crafted with the use of "else" and similar words. In these statements, the use of a semicolon introduces a new conditional clause, known as a nested conditional statement. For example:

- If the collected data indicate the process is in control, continue taking data; otherwise, alert the operator.
- If the water temperature is at or less than 10 degrees Celsius, turn on the heater; or else if the water temperature is at or greater than 80 degrees Celsius, turn on the chiller; otherwise, take no action.

Single Conditional Statements

In MATLAB, a single conditional statement involves the `if` and `end` commands. If two distinct actions must occur as a result of a condition, the programmer can use the `else` command to separate the two actions. The basic structure of a single `if` statement is as follows:

```
if Logical_Expression
        Actions if true
else
        Actions if false
end
```

Note that the *Logical_Expression*, as shown in the structure of a single `if` statement, is a logical expression as described previously.

EXAMPLE 19-7

Write a MATLAB statement to represent the following conditional statement.

If the water velocity (`v`) is less than 10 meters per second, determine the friction factor (`f`) of the piping system using laminar flow given by 64 divided by the Reynolds Number (`Re`).

English (Pseudocode)	MATLAB Code
If the water velocity is slow enough, *use an equation for laminar flow*	`if v < 10` `f = 64/Re`
(period)	`end`

In this example, the comma indicates the separation of the condition and the action that is to be taken if the condition is true. The period marks the end of the statement.

EXAMPLE 19-8

Write a MATLAB statement to represent the following conditional statement.

If the speed of the vehicle is greater than 65 miles per hour, display "Speeding" to the Command Window, otherwise; display "OK."

English (Pseudocode)	MATLAB Code
If the speed is greater than 65, *display "Speeding"*	`if Speed > 65` `fprintf('Speeding')`
Otherwise, *display "OK"*	`else` `fprintf('OK')`
(period)	`end`

In this example, the comma indicates the separation of the condition and the action that is to be taken if the condition is true. The semicolon after the word "otherwise" marks the beginning of the false action. The period marks the end of the statement.

Nested Conditional Statements

If more than two outcomes exist, the program will use an `if-elseif-else` command structure. This is similar to using the nested IF statements in Excel. Each time the word "`if`" appears, whether alone or as part of the `elseif` command, the program must ask a true/false question.

As a minimum, `if` statements can stand alone, without an `else` as seen in Example 19-7 above. As a maximum, for each outcome, the program requires one less `if` statement than outcomes. For example, if you have four desired outcomes, the program will require three IF questions (`if-elseif-elseif`) and one else statement to determine the result.

EXAMPLE 19-9

Write a code segment to generate the menu shown and display the choice of the user in a sentence like: "You bought a Coke." Assume the result of clicking an option in the menu is saved in a variable called `soda` in the workspace. You can assume the user does not close the menu, but makes a valid section from the choices listed.

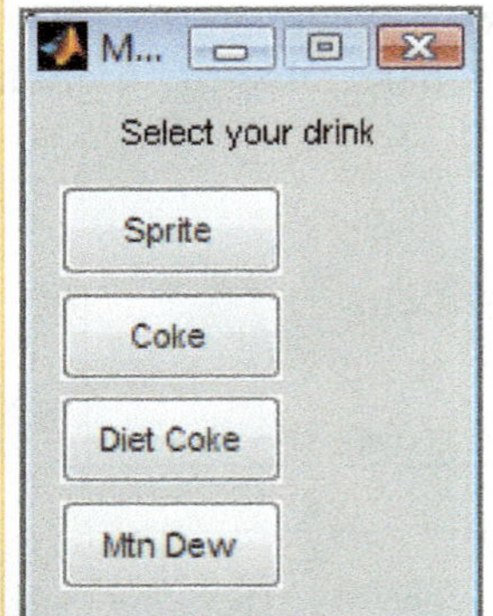

Recall from our discussion of the menu function, the result from a `menu` *selection by the user is the ordinal value of the choice. For example, if the user selects "Coke" from the menu shown, the variable* `soda` *would store the value 2.*

To decipher the choices into text, we can use an IF-ELSEIF-ELSE sequence:

```
soda = menu ('Select your drink','Sprite','Coke','Diet Coke','Mtn Dew');
    if soda == 1
           fprintf('You bought a Sprite\n');
    elseif soda == 2
           fprintf('You bought a Coke\n');
    elseif soda == 3
           fprintf('You bought a Diet Coke\n');
    else
           fprintf('You bought a Mtn Dew\n');
    end
```

COMPREHENSION CHECK 19-4

Write a MATLAB statement to represent the following conditional statement.

If the water temperature (stored in variable `TW`) is at or less than zero degrees Celsius, the phase (stored in the text variable `phase`) is solid; otherwise, if the water temperature is at or greater than 100 degrees Celsius, the phase is vapor; otherwise, the phase is liquid.

COMPREHENSION CHECK 19-5

Write a MATLAB function named `SumItUp` that will accept 3 input variables, and return 2 output variables. If the sum of the first two input variables is greater than 100, the function should display the inputs showing two decimal places in the following format. Assume in this formatting example that the numbers 55, 66 and 77 are the input variables:

```
The inputs are 55.00, 66.00 and 77.00.
```

Otherwise, if the sum of the first two input variables is less than or equal to 100, the function should display the formatted output using no decimal places:

```
The inputs are 25, 10 and 13.
```

The first variable the function returns should be the sum of the first and second variables. The second variable the function returns should be the sum of the second and third variables.

Equivalent Forms of Logic

Conditional statements are similar to traditional language in another way—there are many ways to express the same concept. The order, hierarchy (nesting), and choice of operators are flexible enough to allow the logic statement to be expressed in the way that makes the most sense. All of the following logic statements are equivalent for integer values of `NumTeamMembers`, but all read differently in English:

<table>
<tr><th>Conditional Statement</th><th>English Translation</th></tr>
<tr><td><pre>if NumTeamMembers>=4
  if NumTeamMembers<=5
    fprintf('Team size OK.')
  end
end</pre></td><td>If the number of team members is 4 or more and if the number of team members is 5 or fewer, the team size is okay.</td></tr>
<tr><td><pre>if NumTeamMembers>=4&&NumTeamMembers<=5
  fprintf('Team size OK.')
end</pre></td><td>If the number of team members is 4 or more and 5 or fewer, the team size is okay.</td></tr>
<tr><td><pre>if NumTeamMembers==4||NumTeamMembers==5
  fprintf('Team size OK.')
end</pre></td><td>If the number of team members is 4 or 5 the team size is okay.</td></tr>
</table>

COMPREHENSION CHECK 19-6

Which of the following expressions are equivalent to those shown above to determine if the status of `NumTeamMembers` is acceptable or needs adjusted?

(a)
```
if NumTeamMembers < 4 || NumTeamMembers ~= 5
fprintf('Adjust team size.')
end
```

(b)
```
if NumTeamMembers >= 4 || NumTeamMembers <= 5
fprintf('Team size OK.')
end
```

(c)
```
if NumTeamMembers ~= 4 && NumTeamMembers ~= 5
fprintf('Adjust team size.')
end
```

(d)
```
if NumTeamMembers == 4 && NumTeamMembers == 5
fprintf('Team size OK.')
end
```

19.4 SWITCH STATEMENTS

LEARN TO: **Create a `switch` statement in MATLAB code**
Determine when using a `switch` statement is appropriate
Create an "otherwise" case to execute code when logic isn't matched by the switch

A `switch-case` statement is another way to express `if-elseif-else` statements that test for equality. For certain types of questions, the `switch-case` statement is perhaps a little simpler to write and to understand.

The most important point about `switch-case` is that it ONLY tests for equality. If you need to make other types of relational tests (such as greater than), you should use `if` statements. Also, except in a very limited context, you cannot use logical operations (and, or, not) with `switch-case` statements.

The format of the `switch-case` structure is:

```
switch switch_expression
  case case_expression_1
       Execute code here if switch_expression==case _expression_1
  case case_expression_2
       Execute code here if switch_expression==case _expression_2
      .
      .
      .
  case case_expression_N
       Execute code here if switch_expression==case _expression_N
  otherwise
       Execute code here if switch_expression does not equal
       any of the cases
end
```

In general, the `switch-case` statement compares the value of a `switch` expression, with different `case` expressions. When it finds a `case` expression that equals the `switch` expression, it executes the code associated with that case, then goes to the end of the `switch` structure and continues. Note that only one case will be executed (the first one that is equal) each time the switch case is executed.

A common use of `switch-case` statements is processing menu selections.

EXAMPLE 19-10 Write a program that allows the user to enter an angle in degrees and then select a trigonometric function from a menu to calculate. Express the result in a sentence, such as:

```
The sine of 30 degrees is 0.5000
```

One possible code:

```
Angle = input ('Enter angle in degrees: \n');
radAngle = Angle*pi/180;
trig = menu ('Select a function','sine','cosine','tangent');

switch trig
   case 1
      result = sin (radAngle);
      fprintf('The sine of %0.0f is %0.4f.\n',radAngle,result)
   case 2
      result = cos (radAngle);
      fprintf('The cosine of %0.0f is %0.4f.\n',radAngle,result)
   case 3
      result = tan (radAngle);
      fprintf('The tangent of %0.0f is %0.4f.\n',radAngle,result)
   otherwise
      fprintf('No selection made. Result set to NaN.\n')
      result = NaN;
end
```

If button number 2 (cosine) is selected, the cosine of the angle is displayed. Note that if the user closes the menu, the code between `otherwise` *and* `end` *is executed and the result is set to* `NaN` *(not a number).*

The expressions do not have to be simple variables or constants: they may also include calculations.

EXAMPLE 19-11

Write a code segment to accept from the user the nominal diameter of a bolt and the actual diameter of a specific bolt. The program should then classify the bolt diameter as within less than 1%, 2%, or 5%, of the nominal value, as well as indicating if the bolt diameter is too far out of specifications when the nominal and actual diameters differ by 5% or more.

One possible code:

```
NomDiam = input ('Enter nominal diameter of bolt: \n');
BoltDiam = input ('Enter actual diameter of bolt: \n');
percent = BoltDiam/NomDiam*100-100; % Calculate error

switch abs(fix(percent)) % round toward zero and make positive
   case 0
      fprintf('Diameter within < 1% of desired value.\n')
   case 1
      fprintf('Diameter between 1% and < 2% of desired value.\n')
   case {2,3,4}
      fprintf('Diameter between 2% and < 5% of desired value.\n')
   otherwise
      fprintf('Out of specifications (>= 5% error).\n')
end
```

The third case bears a bit of explanation. You can place multiple expressions (in this example, simple constants) in a cell array as the case expression. If ***any*** *of the expressions in the cell array are equal to the* `switch` *expression, then the instructions associated with that* `case` *will be executed unless, of course, a prior case was true. This is the "limited use of logical operations" mentioned earlier, since it effectively implements an "or", although the "or" is not explicitly coded.*

Text strings can also be used in `switch-case` structures. Note that this is case-sensitive, since it effectively implements a `strcmp`, not a `strcmpi`. If you need to test strings where case does not matter, it is probably better to use `if` statements in conjunction with `strcmpi`.

EXAMPLE 19-12

Write a code segment to allow the user to enter their favorite type of pet, then the program will describe the animal as the output displayed to the user. If the user types a value that isn't one of the values specified, the code indicates that the user entry doesn't match any of the recognized animals.

One possible code:

```
pet = input ('What is your favorite pet? (Bird, Cat, or Dog)\n','s');

switch pet
   case 'Bird'
      fprintf('Your favorite pet flies through the air.\n')
   case 'Cat'
      fprintf('Your favorite pet makes a yowling noise. \n')
```

```
    case 'Dog'
        fprintf('Your favorite pet barks. \n')
    otherwise
        fprintf('Your entry does not match my database. \n')
end
```

Note that in this example, the first letter of the response must be capitalized to have one of the first three cases selected.

COMPREHENSION CHECK 19-7

Write a MATLAB code using `switch-case` to allow the user to choose from a menu the type of water sample they wish to prepare. Based on their choice, the program should provide guidelines for how to prepare the sample. For example, if the user selects "solid", the program should state:

```
For a solid, water must be at a temperature
less than 0 degrees Celsius.
```

19.5 ERRORS AND WARNINGS

LEARN TO: **Create a warning statement to display in the Command Window**
Create an error expression to display in the Command Window
Design logic using `try-catch` statements to gracefully terminate the code

MATLAB contains built-in functions that allow you to create custom error and warning messages within your code, similar to those you encounter when you have a syntax error in your code or encounter a runtime error. You can use these functions, in conjunction with logical operators, to build error checking and prevention into your MATLAB code.

Error Messages

You might have noticed that when you run MATLAB code with syntax errors, the Command Window will display errors in red letters to report the error message rather than the standard black text font. In MATLAB, you can actually create your own custom error messages.

Error messages in MATLAB are shown in the Command Window in red letters:

```
Undefined function or variable 'A'.
```

If you would like to build in your own error checking in MATLAB, you can actually program custom error messages that may be more helpful than the default messages displayed by MATLAB using the error function.

EXAMPLE 19-13

We are given the weight (in variable w) in newtons of an object as well as the mass (in variable m) in kilograms and want to determine the gravity in meters per second squared. To solve this, we will use the equation $g = w/m$, but we want to wrap that expression inside a conditional just in case the user provides a zero mass, which does not make sense in our situation, but will not cause an error in MATLAB—if you divide a number by 0 in MATLAB, the value "Inf" is calculated, which represents infinity.

```
>> w=530;
>> m=0;
>> g=w/m
g=
   Inf
```

Instead, we could write the following segment of code to force MATLAB to throw an error message:

```
if m==0
    error('Error: The mass cannot be zero!');
else
      g=w/m;
end
fprintf('The gravity is %0.1f m/s^2\n',g);
```

When you run this code given the w and m above, the message below will display in the Command Window and your code will stop executing.

```
Error: The mass cannot be zero!
```

Warning Messages

If you want to provide a warning to the user but do not want MATLAB to stop running, you can use the **warning** function. This will allow the program to continue, while giving the user information about potential issues in orange letters in the Command Window.

EXAMPLE 19-14

Repeat Example 19-13, but this time if the user sets the mass variable to be zero, we want to display a warning message to the user and re-define the mass to be 3 kilograms to compute gravity.

```
w=530;
m=0;
if m==0
    warning('Mass=0! Using 3 kg for the mass');
    m=3;
end
g=w/m;
fprintf('The gravity is %0.1f m/s^2.\n',g);
```

*When you run this code, the warning message will display in the Command Window and your code will continue executing. Using the **warning** function, the **else** is unnecessary since the calculation will be done in either case.*

```
Warning: Mass=0! Using 3 kg for the mass
The gravity is 176.7 m/s^2.
```

Try-Catch Statements

To customize how MATLAB handles reporting error messages, you can use try-catch statements to handle error exceptions. An exception, in MATLAB, is a structure that contains special information such as a description of the error, as well as information concerning where the error occurred in a particular file. The try-catch statement is formatted, as follows:

```
try
        MATLAB code...
        goes here...
        .
        .
        .
catch e
        MATLAB code that executes on error occurs here.
end
```

Try-catch statements can be used over large blocks of code, even entire programs or functions, but their real value comes into play when used on smaller blocks of code to provide more meaningful feedback to the user running your program.

EXAMPLE 19-15

Write a code segment to ask the user to enter a matrix. If the matrix can be muliplied by a 2×4 matrix, the program will exectue the operation and display "Finished!" when complete. If the multiplication is invalid, the program will produce an error message.

```
try
   i = input ('Type a matrix: \n');
   m = i * [3,3,3,3;4,4,4,4];
   fprintf('Finished!\n')
catch e
   error('Houston, we have a problem:\n\n %s', e.message)
end
```

Sample Output:

```
Type a matrix: 3
Finished!
```

In the output above, there were no errors, so the catch statements did not execute.

```
Type a matrix: [1 1 2 3 5; 2 2 1 3 4; 3 3 3 3 3]
Houston, we have a problem:
Inner matrix dimensions must agree.
```

In the output above, there was an error, so the code terminated and the code in the error handler (the `catch` *block) executed. Note the* `fprintf` *in the* `try` *block did NOT execute.*

In-Class Activities

ICA 19-1

Answer the following questions.

(a) For what integer values of D will the code between if and end execute?

```
if D >= -1 && 3 > D
        code
end
```

(b) For what values of F will the code between if and end execute?

```
if F >= 100 || F <= 5
        code
end
```

(c) What combinations of values for B and Q will cause a warning to be displayed?

```
if B>100 || (B<20 && Q~=0)
        fprintf('WARNING! Parameters out of bounds.\n')
end
```

ICA 19-2

Answer the following questions.

(a) For what integer values of G will the code between if and end execute?

```
if G < 2 && -5 <= G
        code
end
```

(b) For what values of R will the code between if and end execute?

```
if R < 75 || R >= 95
        code
end
```

(c) What combinations of values for W and M will cause a warning to be displayed?

```
if W<50 || (W>120 && M~=0)
        fprintf('WARNING! Parameters out of bounds.\n')
end
```

ICA 19-3

For each task listed, write a single MATLAB statement that will accomplish the stated goal.

(a) Replace each negative number in matrix M1 with zero.

(b) Given that a scalar value N is already defined, determine the sum of all elements in vector V1 that are greater than N. *HINT: Use the sum function.*

ICA 19-4

For each task listed, write a single MATLAB statement that will accomplish the stated goal.

(a) Replace each negative number in matrix `M2` with −9999.
(b) Given that a scalar value `N` is already defined, determine the sum of all elements in matrix `M3` that are less than or equal to `N`. *HINT: Use the sum function.*

ICA 19-5

What is stored in variable `A` after the following code segment is executed? If the program returns an error, indicate why.

(a)
```
A=5;
if A/2>2
  A = A*-1;
else
  A = A/2;
end
```

(b)
```
x=2; y=4; A=0;
if x+y<4
  A = A+2;
elseif x+y<6
  A = A+4;
end
```

(c)
```
T=6; A=-1;
if T/3>2
  A = A*-1;
elseif T/3<=2
  A = A+2;
end
```

(d)
```
x=2;y=0;A=10;
if x<=0&&y<=0
  A = A+2;
elseif x>=0&&y<0
  A = A/2;
elseif x>=2&&y<=0
  A = A*2;
end
```

ICA 19-6

What will be displayed by the following code in each of the cases listed below?

```
if flag==1 && alt>=30000
        fprintf('Normal operation at %0.0f feet.\n',alt)
elseif flag==0 || alt==0
        fprintf( On Ground')
elseif flag==2 && alt < 30000
        fprintf('Currently at %0.0f feet and climbing\n',alt)
elseif flag==3
        fprintf('Currently at %0.0f feet and descending\n',alt)
else
        fprintf('Status transitional')
end
```

(a) `flag` = 1; `alt` = 25000
(b) `flag` = 0; `alt` = 7500
(c) `flag` = 2; `alt` = 10000
(d) `flag` = 1; `alt` = 37000
(e) `flag` = 3; `alt` = 35000

ICA 19-7

A menu is generated using the following code:

```
Status=menu('ClassStanding','Freshman','Sophomore','Junior','Senior');
```

Write a segment of code using `if-elseif-else` that will classify the person making a menu selection as a new student (freshman) or a continuing student (sophomore, junior, or senior). The code should display one of the following messages:

```
You are a new student.
You are a continuing student.
You did not make a selection.
```

ICA 19-8

A menu is generated using the following code:

```
Status=menu('ClassStanding','Freshman','Sophomore','Junior','Senior');
```

Write a segment of code using `switch-case` that will classify the person making a menu selection as a new student (freshman) or a continuing student (sophomore, junior, or senior). The code should display one of the following messages:

```
You are a new student.
You are a continuing student.
You did not make a selection.
```

ICA 19-9

Write a program using `if-elseif-else` statements that displays a menu of traffic light colors (green, yellow, red). Depending on which button is pressed, the user should then be told to continue, slow, or stop. If the user closes the menu without making a selection, the user should be told to stop.

ICA 19-10

Write a program using `switch-case` statements that displays a menu of traffic light colors (green, yellow, red). Depending on which button is pressed, the user should then be told to continue, slow, or stop. If the user closes the menu without making a selection, the user should be told to stop.

ICA 19-11

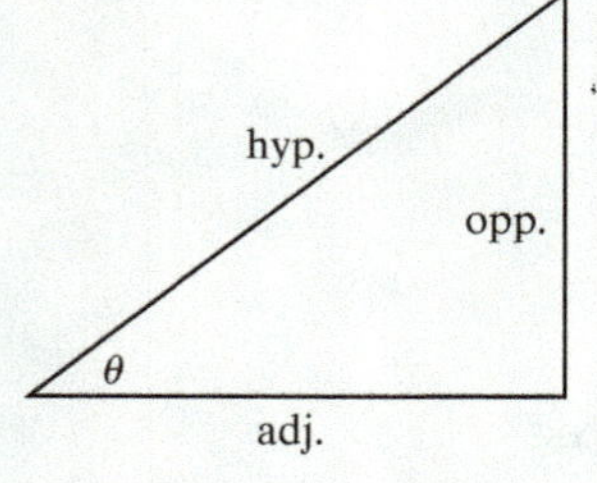

Write a program that asks the user to enter the length of the hypotenuse, opposite, and adjacent sides of a right triangle with respect to an angle (θ). After the user has entered all three values, a menu should pop up asking if the user wants to calculate sin(θ), cos(θ), tan(θ), cot(θ), cosec(θ), or sec(θ). The program should display a final message, such as "`For a triangle of sides h, o and a, sin(θ) = 0.3,`" where h, o, and a are replaced by the actual lengths entered by the user, displayed to one decimal place. If the user enters lengths that do not work with a right triangle, an error message should appear and the program should terminate. If the user closes the menu instead of making a valid selection, an appropriate error message should appear and the program should terminate. You may use `if` statements and/or `switch-case` statements as appropriate.

ICA 19-12

We go to a state-of-the-art amusement park. All the rides in this amusement park contain biometric sensors that measure data about potential riders while they are standing in line. Assume the sensors can detect a rider's age, height, weight, heart problems, and possible pregnancy. Help the engineers write the conditional statement for each ride at the park based on their safety specifications.

	Variable Definitions:
A	% Age of the potential rider (as an integer)
H	% Height of the potential rider (as an integer)
HC	% Heart Condition status ('yes' if the person has a heart condition, 'no' otherwise)
P	% Pregnancy status (1 if pregnant, 0 otherwise)

(a) The Spinning Beast: All ricers must be 17 years or older and more than 62 inches tall and must not be pregnant or have a heart condition.

```
if ______________________________________
  fprintf ('Sorry, you cannot ride this ride');
end
```

(b) The Lame Train: All riders must be 8 years or younger and must not be taller than 40 inches.

```
if ______________________________________
  fprintf ('Sorry, you cannot ride this ride');
end
```

(c) The MATLAB House of Horror: All riders must be 17 years or older and must not have a heart condition.

```
if ______________________________
  fprintf ('Welcome to the MATLAB House of Horror!');
end
```

(d) The Neck Snapper: All riders must be 16 years or older and more than 65 inches tall and must not be pregnant or have a heart condition.

```
if ______________________________________
  fprintf ('Welcome to The Neck Snapper!');
end
```

(e) The Bouncy Bunny: All riders must be between the ages of 3 and 6 (including those 3 and 6 years old)

```
if ______________________________________
  fprintf ('This ride is made just for you!');
end
```

ICA 19-13

A phase diagram for carbon and platinum is shown. If it is assumed the lines shown are linear, the mixture has the following characteristics. The endpoints of the division line between these two phases are labeled on the diagram.

- Below 1,700 degrees Celsius, it is a mixture of solid platinum (Pt) and graphite (G).
- Above 1,700 degrees Celsius, there are two possible phases: a Liquid (L) phase and a Liquid (L) + Graphite phase (G).

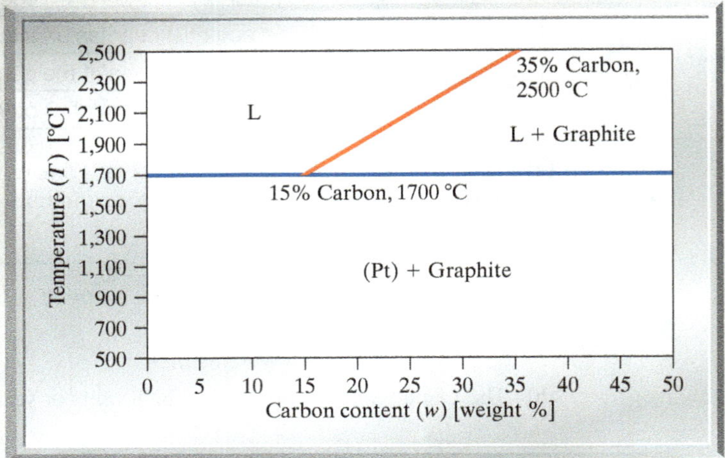

Write a program to determine the phase. The program should ask the user to enter the weight percent carbon and the temperature in degrees Celsuis. Call the phases "Pt + G," "L," and "L + G," for simplicity. If the point falls directly on the $T = 1700$ deg C line, include the point in the "Pt + G" phase. If the point falls directly on the L, L + G line, include the point in the "L + G" phase.

Store the phase as text in a variable. The equation of the line dividing the L and L + G phases must be found in the program using `polyfit`. The program should produce a formatted output statement to the command window, similar to "`For X.XX weight percent carbon and a temperature of YYY degrees Celsius, the phase is PHASE,`" where X.XX, YYY and Phase are replaced by the actual values formatted as shown.

The partial code below is designed to implement this program. You are to fill in the missing sections of code as appropriate to complete the program.

```
[Q1.                ]        % Appropriate housekeeping commands
w = input ('Type the Carbon content [weight %]: ');
T = input ('Type the temperature [deg C]: ');

% set parameters for L / L + G line
Carbon1 = [[Q2.  ]];
Temp1 = [[Q3.  ]];

% set parameters for L / Pt + G and L + G / Pt + G line
Carbon2 = [[Q4.  ] ];
Temp2 = [ [Q5.  ]];

% determine polyfit parameters for L / L+G line
PhaseLine = polyfit([Q6.              ] );
TPhaseLine = [Q7.              ] );
```

```
    if [Q8.              ]            % Add appropriate question here
            Phase = 'Pt + G';
    elseif [Q9.          ]            % Is it a liquid?
         [Q10.          ]             % Add appropriate code here
[Q11.              ]        % First of two lines to complete if statement
[Q12.              ]        % Second of two lines to complete if statement
end

fprintf [Q13          ]     % Add appropriate code
```

ICA 19-14

The graph shows a phase diagram for lead–tin solder. The most common alloy available commercially is 60% Sn (tin) and 40% Pb (lead), although eutectic solder (63% Sn and 37% Pb) is easily obtained and has some distinct advantages, primarily related to the lack of a "pasty" range, the range of temperatures in which the alloy is a mixture of solid within a liquid as it cools from liquid to solid. The gridlines have been removed from this graph to make it easier to read the phase locations.

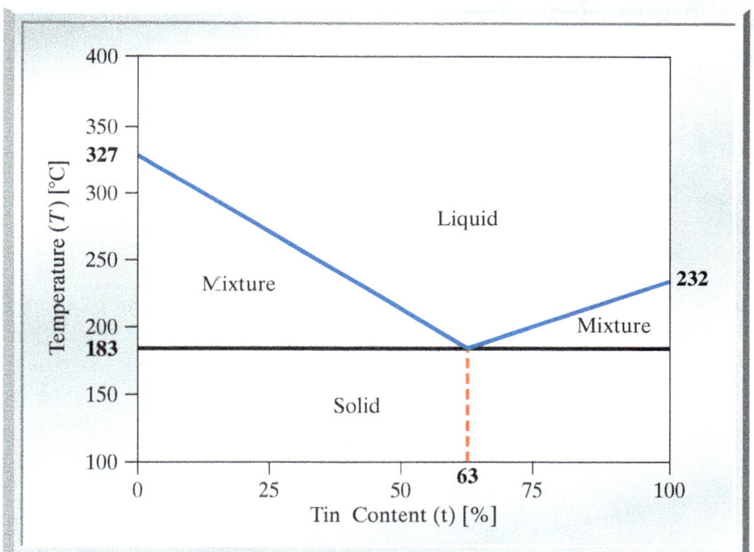

A function is to be written that will classify the phase of a eutectic Sn–Pb alloy given the percentage of tin and the temperature in units of degrees Celsius. The function should be named `SnPbPhase` and should accept two scalars:

- `Temp`, containing the temperature in °C
- `PercSn`, the percentage of tin

The function should return a text string named `Phase` containing one of the following four values:

- Liquid
- Solid
- Mixture
- Eutectic point

If a specific combination of percentage and temperature is exactly on one of the lines (other than at the intersection of the eutectic point at 63% and 183°C), the classification should be with the region below the line. The equation of the lines dividing the phases must be found in the program using `polyfit`. The program should produce a formatted output statement to the command window, similar to "`For X.XX percentage of tin and a temperature of YYY deg C, the phase is PHASE`," where X.XX, YYY and Phase are replaced by the actual values formatted as shown.

The partial code below is designed to implement this function. You are to fill in the missing sections of code as appropriate to complete the function.

```
[Q1.          ]  % Declaration of function and passed parameters

% set parameters for Mixture / Liquid line when Tin < 63%
Tin1=[[Q2. ]];
Temp1=[[Q3. ]];
LoSn = polyfit([Q4.            ]);
TempLoSn = [Q5.          ]% Temperature on LoSn line at given percent Tin
% set parameters for Mixture / Liquid line when Tin > 63%
Tin2=[ [Q6.  ]];
Temp2=[[Q7.  ] ];
HiSn = polyfit( [Q8.            ]);
TempHiSn = [Q9.            ]); % Temperature on HiSn line at given percent Tin

if [Q10.          ]      % Add appropriate question here
  Phase = 'Eutectic Point';
elseif [Q11.            ]  % Is it solid?
  [Q12.            ]       % Add appropriate code here
elseif PercSn<=63 && [Q13.            ]  % Complete the question
  Phase='Liquid';
elseif [Q14.        ]      % Add appropriate question here
  Phase='Liquid';
[Q15.            ]         % First of two lines to complete if statement
[Q16.            ]         % Second of two lines to complete if statement
end
fprintf [Q17.        ]     % Add appropriate code
```

Upon reflection, you realize that the second and third `elseif` statements (the two that classify the material as liquid) can be written as a single `elseif` statement using a question with ONLY a SINGLE logical operator ("`and`" or "`or`"). What is the correct question to ask in a single `elseif` statement to correctly classify the material as liquid?

```
elseif [Q18.          ]   % Add appropriate question here
Phase='Liquid';
```

Chapter 19 REVIEW QUESTIONS

1. The specific gravity of gold is 19.3. Write a MATLAB program that will ask the user to input the mass of a cube of solid gold in units of kilograms and display the length of a single side of the cube in units of inches. The output should display a sentence like the one shown below, with the length formatted to 2 decimal places. If the user types a negative number or zero for the mass of the cube, your program should display an error message and terminate.

   Sample Input / Output (multiple scenarios):

   ```
   Enter the mass of the cube [kilograms]: -3
   Error: Mass must be greater than zero grams.
   ```

   ```
   Enter the mass of the cube [kilograms]: 0.4
   The length of one side of the cube is 1.08 inches.
   ```

2. An unmanned X-43 A scramjet test vehicle has achieved a maximum speed of Mach number 9.68 in a test flight over the Pacific Ocean. Mach number is defined as the speed of an object divided by the speed of sound. Assuming the speed of sound is 343 meters per second, write a MATLAB program to determine speed in units of miles per hour. Your program should ask the user to provide the speed as Mach number and return the speed in miles per hour in a formatted sentence, displayed as an integer value, as shown in the sample output below. If the user provides a negative value for the Mach number, your program should display an error message and terminate.

   Sample Input / Output (multiple scenarios):

   ```
   Enter the speed as a Mach number: -2
   Error: The Mach number must not be negative.
   ```

   ```
   Enter the speed as a Mach number: 0
   The speed of the plane is 0 mph.
   ```

   ```
   Enter the speed as a Mach number: 9.68
   The speed of the plane is 7425 mph.
   ```

3. A rod on the surface of Jupiter's moon Callisto has a volume measured in cubic meters. Write a MATLAB program that will ask the user to type in the volume in cubic meters in order to determine the weight of the rod in units of pounds-force. The specific gravity is 4.7. Gravitational acceleration on Callisto is 1.25 meters per second squared. The input and output of the program should look similar to the output below. Be sure to report the weight as an integer value. If the user types a negative number or a number greater than 100 cubic meters for the volume of the rod, your program should display an error message using the `error` function indicating that the provided input is outside of the desired range and terminate.

   Sample Input / Output (multiple scenarios):

   ```
   Enter the volume of the rod [cubic meters]: -10
   Error: Volume must be between 0 and 500 cubic meters
   ```

   ```
   Enter the volume of the rod [cubic meters]: 501
   Error: Volume must be between 0 and 500 cubic meters
   ```

   ```
   Enter the volume of the rod [cubic meters]: 0.3
   The weight of the rod is 397 pounds-force.
   ```

4. The Eco-Marathon is an annual competition sponsored by Shell Oil, in which participants build special vehicles to achieve the highest possible fuel efficiency. The Eco-Marathon is held around the world with events in the United Kingdom, Finland, France, Holland, Japan, and the United States.

A world record was set in the Eco-Marathon by a French team in 2003 called Microjoule with a performance of 10,705 miles per gallon. The Microjoule runs on ethanol. Write a MATLAB program to determine how far the Microjoule will travel in kilometers given a user-specified amount of ethanol, provided in units of grams. Your program should ask for the mass using an input statement and display the distance in a formatted sentence similar to the output shown below. If the user provides a mass less than zero or greater than 500 grams, your program should display an error message using the `error` function indicating that the provided mass of ethanol is outside of the desired range of input.

Sample Input / Output: (multiple scenarios):

```
Enter mass of ethanol [grams]: -15
Error: Mass must be between 0 and 500 grams
```

```
Enter mass of ethanol [grams]: 1000
Error: Mass must be between 0 and 500 grams
```

```
Enter mass of ethanol [grams]: 100
The distance the Microjoule traveled is 577 kilometers.
```

5. Assume a variable *R* contains a single number. Write a short piece of MATLAB code that will:

- Generate the message: `The square root of XXXX is YYYY`, if *R* is nonnegative.
- If *R* is negative, the code should generate the message: `R is negative. The square root of XXXX is YYYYi.`

The value of *R* should be substituted for XXXX, and the value of the square root of the magnitude of *R* should be substituted for YYYY. Both numbers should be displayed with three decimal places.

NOTE

Due to several complicated nonlinear losses in the system that are far beyond the scope of this course, this is a case of a model in which the exponent does not come out to be an integer or simple fraction, so rounding to two significant figures is appropriate. In fact, this model is only a first approximation—a really accurate model would be considerably more complicated.

6. Write a program that gathers two data pairs in a 2×2 matrix and then asks the user to input another value of *x* for which a value of *y* will be interpolated or extrapolated. Report the value of *y* in the correct sentence below (with numbers in the blanks).

```
Given x = _______, interpolation finds that y = _______ or
Given x = _______, extrapolation finds that y = _______.
```

Sample Input/Output Matrix [0 10, 10 20] is entered and the user chooses 5:

```
Given x = 5, interpolation finds that y = 15.
```

7. Create a program to determine whether a user-specified altitude [meters] is in the troposphere, lower stratosphere, or upper stratosphere. The program should include a check to ensure the user entered a positive value. If a non-positive value is entered, the program should inform the user of the error and terminate. If a positive value is entered, the program should calculate and report the resulting temperature in units of degrees Celsius [°C] and pressure in units of kilopascals [kPa]. Refer to the atmosphere model provided by NASA: http://www.grc.nasa.gov/WWW/K-12/airplane/atmosmet.html. **Sample Input/Output** Altitude = 500 meters.

```
An altitude of 500 is in the troposphere with a temperature of
12 degrees C and pressure of 96kPa.
```

8. Create a program to determine whether a given Mach number is subsonic, transonic, supersonic, or hypersonic. Assume the user will enter the speed of the object, and the program will determine the Mach number. As output, the program will display the Mach rating. The program should include a check to ensure the user entered a positive value. If a non-positive value is entered, the program should inform the user of the error and terminate.

Refer to the NASA page on Mach number: http://www.grc.nasa.gov/WWW/K-12/airplane/mach.html. **Sample Input/Output** Speed = 100 m/s.

```
Subsonic, Mach number is 0.3.
```

9. Humans can see electromagnetic radiation when the wavelength is within the spectrum of visible light. Create a program to determine if a user-specified wavelength [nanometer, nm] is one of the six spectral colors listed in the chart below. Your program should ask the user to enter a wavelength, then indicate in which spectral color the given wavelength falls or state if it is not within the visible spectrum.

Color	Wavelength Interval
Red	~ 700–635 nm
Orange	~ 635–590 nm
Yellow	~ 590–560 nm
Green	~ 560–490 nm
Blue	~ 490–450 nm
Violet	~ 450–400 nm

10. For the protection of both the operator of a zero-turn radius mower and the mower itself, several safety interlocks must be implemented. These interlocks and the variables that will represent them are listed below.

Interlock Description	Variable Used	State
Brake Switch	Brake	True if brake on
Operator Seat Switch	Seat	True if operator seated
Blade Power Switch	Blades	True if blades turning
Left Guide Bar Neutral Switch	LeftNeutral	True if in neutral
Right Guide Bar Neutral Switch	RightNeutral	True if in neutral
Ignition Switch	Ignition	True if in run position
Motor Power Interlock	Motor	True if motor enabled

For the motor to be enabled (thus capable of running), all the following conditions must be true.

- The ignition switch must be set to "Run."
- If the operator is not properly seated, both guide levers must be in the locked neutral position.
- If either guide lever is not in the locked neutral position, the brake must be off.
- If the blades are powered, the operator must be properly seated.

Write a function that will accept all of the above variables except `Motor`, decide whether the motor should be enabled or not, and place true or false in `Motor` as the returned variable.

Sample Input/Output If the parameters are set to 0,0,0,0,0,1, the Motor output will be 0.

```
If the parameters are set to 0,1,0,1,1,1, the Motor output will
be 1.
```

11. Most resistors are so small that the actual value would be difficult to read if printed on the resistor. Instead, colored bands denote the value of resistance in ohms. Anyone involved in constructing electronic circuits must become familiar with the color code, and with practice, one can tell at a glance what value a specific set of colors means. For the novice, however, trying to read color codes can be a bit challenging.

You are to design a program that, when a user enters a resistance value, the program will display (as text) the color bands in the order they will appear on the resistor.

The resistance will be entered in two parts: the first two digits, and a power of 10 by which those digits will be multiplied. The user should be able to select each part from a menu. You should assume the ONLY values that can be selected by the user are the ones listed in Tables 19-4 and 19-5.

As examples, a resistance of 4,700 ohms has first digit 4 (yellow), second digit 7 (violet), and 2 zeros following (red). A resistance of 56 ohms would be 5 (green), 6 (blue), and 0 zeros (black); 1,000,000 ohms is 1 (brown), 0 (black), and 5 zeros (green). There are numerous explanations of the color code on the web if you want further information or examples.

Table 19-3 Standard resistor values

10	12	15	18	22	27	33	39	47	56	68	82

Table 19-4 Standard multipliers

1	10	100	1000	10000	100000	1000000

Table 19-5 Color codes

0	1	2	3	4	5	6	7	8	9
Black	Brown	Red	Orange	Yellow	Green	Blue	Violet	Grey	White

12. A phase diagram (from http://www.eng.ox.ac.uk/~ftgamk/engall_tu.pdf) for copper–nickel shows that three phases are possible: a solid alpha phase (a solid solution), a liquid solution, and a solid–liquid combination.

Write a program that takes as input a mass percent of nickel and a temperature in units of degress Celsius and gives as output a message indicating what phases are present. Assume the region borders are linear, and determine a method for dealing with any points that lie directly on the lines. The equation of the line dividing the phases must be found in the program using `polyfit`. The program should produce a formatted output statement to the command window, similar to "`For w weight percent nickel and a temperature of T degrees Celsius, the phase is PHASE`," where w, T and Phase are replaced by the actual values.

The program should also reproduce the graph as shown here with all phase division lines, and indicate the point entered by the user with a symbol.

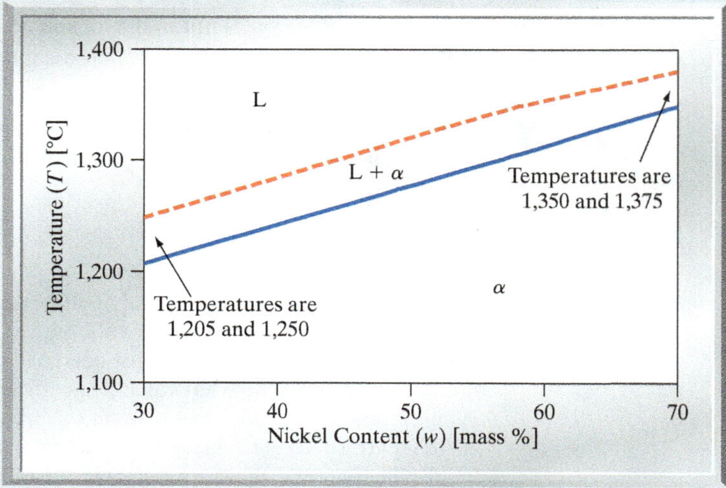

13. This generic phase diagram, based on the temperature and composition of elements A and B, is taken from http://www.soton.ac.uk/~pasr1/build.htm, where a description of how phase diagrams are constructed can also be found. The alpha and beta phases represent solid solutions of B in A and of A in B, respectively.

The eutectic line represents the temperature below which the alloy will become completely solid if it is not in either the alpha or beta region. Below that line, the alloy is a solid mixture of alpha and beta. Above the eutectic line, the mixture is at least partially liquid, with partially solidified lumps of alpha or beta in the labeled regions.

Assume the following:

- The melting point of pure A is 700 degrees Celsius.
- The melting point of pure B is 800 degrees Celsius.
- The eutectic line is at 300 degrees Celsius, and the eutectic point occurs when the composition is 50% B.
- The ends of the eutectic line are at 15% B and 85% B.

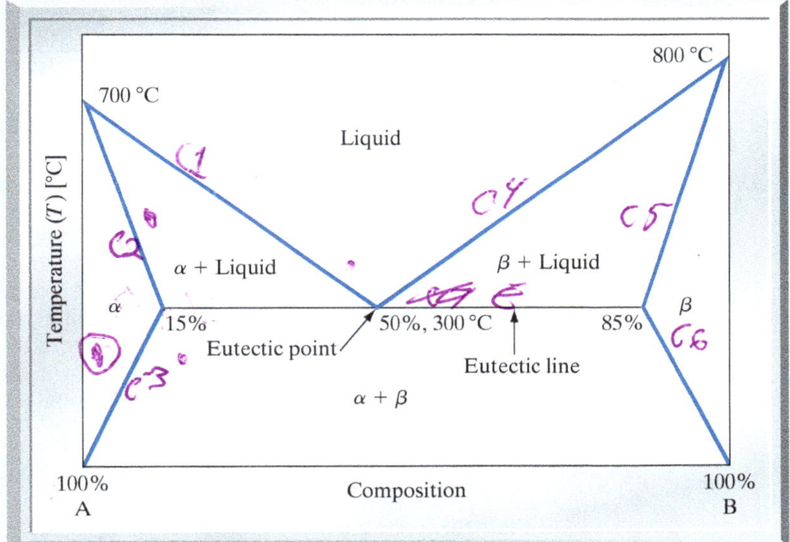

Using the simplified phase diagram, which substitutes straight lines for the curved lines typical of phase diagrams, write a MATLAB program that takes as input the mass percent of B and the temperature in units of degrees Celsius and returns the phases that may exist under those conditions. The equation of the line dividing the phases must be found in the program using `polyfit`. The program should produce a formatted output statement to the command window, similar to "`For the composition of x% A, y% B and a temperature of T degrees Celsius, the phase is PHASE`," where x, y, T and Phase are replaced by the actual values.

The program should also reproduce the graph as shown here with all phase division lines, and indicate the point entered by the user with a symbol.

Your MATLAB program should include special notes if the provided conditions are on the eutectic line or at the eutectic point.

14. Soil texture can be classified with a moist soil sample and the questions in the table below from the Soil Texture Key found at http://www.pasture4horses.com/soils/hand-texturing.php/. Each answer results in either a soil classification or directions to continue. Write a program that implements this decision tree. You may find it useful to download the key itself from the above address. You may also want to compare your results to an online implementation of the key found at the same website. Your program should include the use of menu windows to gather user responses.

Question	Yes	No
1. Does the soil feel or sound noticeably sandy?	Go to Q2	Go to Q6
2. Does the soil lack all cohesion?	SAND	Go to Q3
3. Is it difficult to roll the soil into a ball?	LOAMY SAND	Go to Q4
4. Does the soil feel smooth and silky as well as sandy?	SANDY SILT LOAM	Go to Q5
5. Does the soil mould to form a strong ball that smears without taking a polish?	SANDY CLAY LOAM	SANDY LOAM
6. Does the soil mould to form an easily deformed ball and feel smooth and silky?	SILT LOAM	Go to Q7
7. Does the soil mould to form a strong ball that smears without taking a polish?	Go to Q8	Go to Q10
8. Is the soil also sandy?	SANDY CLAY LOAM	Go to Q9
9. Is the soil also smooth and silky?	SILTY CLAY LOAM	CLAY LOAM
10. Does the soil mould like plasticine, polish, and feel very sticky when wetter?	Go to Q11	UNKNOWN SOIL
11. Is the soil also sandy?	SANDY CLAY	Go to Q12
12. Is the soil also smooth and buttery?	SILTY CLAY	CLAY

15. Your boss hands you the following segment of MATLAB code that implements a safety control system in an automobile. Recently, the engineer who wrote the code below was fired for their inability to write efficient code. The control system below takes 5 minutes for the car to execute because it was not written using nested-`if` statements, and worse, it does not work with multiple sensors because the former employee could not figure out how to make use of nested-`if` statements! Your boss has instructed you to take the former employee's code and fix it, or you might meet the same fate as the original programmer!

```
% Variable Definition:
% Input:
% Ignition: 0 if engine off, 1 if key in ignition, 2 if engine on
% Belt: 1 if buckled, 0 if unbuckled, -1 if sensor broken
% HLamp: 1 if all bulbs ok, 0 if 1 bulb out, -1 if 2 or more out
% TLamp: 1 if all bulbs ok, 0 if 1 bulb out, -1 if 2 or more out
% Light: 1 if sky is bright, 0 if sky is dark (need lamps)
% Output:
% Safety: 1 if safe, 0 if warn/caution, -1 if unsafe

function[Safety]=Vehicle(Ignition,Belt,HLamp,TLamp,Light)

if Ignition==1 || Ignition == 0
      Safety = 1;
end
if Ignition==2 && Belt == 1
      Safety = 1;
end
if Ignition==2 && Belt == 0
      Safety = 0;
end
```

```
if Ignition==2 && Belt == -1
        Safety = -1;
end
if Ignition==2 && Belt == 1 && HLamp == 1
        Safety = 1;
end
if Ignition==2 && Belt == 1 && HLamp == 0
        Safety = 0;
end
if Ignition==2 && Belt == 1 && HLamp == -1
        Safety = 0;
end
if Ignition==2 && Belt == 1 && HLamp == -1 && Light==0
        Safety = -1;
end
if Ignition==2 && Belt == 1 && TLamp == 1
        Safety = 1;
end
if Ignition==2 && Belt == 1 && TLamp == 0
        Safety = 0;
end
if Ignition==2 && Belt == 1 && TLamp == -1
        Safety = 0;
end
if Ignition==2 && Belt == 1 && TLamp == -1 && Light == 0
        Safety = -1;
end
```

16. Assume you are required to generate the menus shown below in a MATLAB program. Write a program to generate these menus and display the choice of the user in a sentence like: "You selected a car with automatic transmission." If the user clicks "Other," an input statement should allow them to type a different vehicle type (as a string). For this program, use `if-elseif` statements instead of `switch` statements.

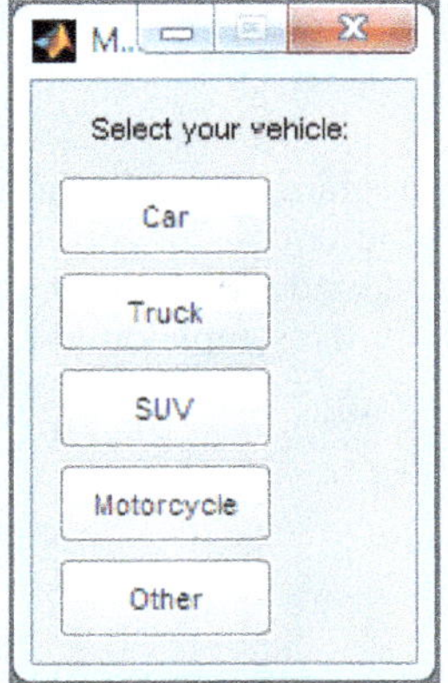

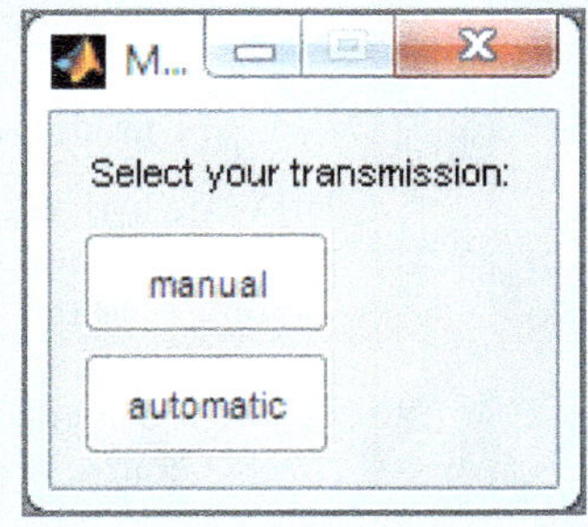

17. Assume you are required to generate the menus shown with Review Question 19-16 in a MATLAB program. Write the program necessary to generate these menus and display the choice of the user in a sentence like "You selected a car with automatic transmission." If the user clicks "Other," an input statement should allow them to type a different vehicle type (as a string). For this program, use `switch` statements instead of `if-elseif` statements.

18. Assume you are required to generate the menus shown below in a MATLAB program. Write a program to generate these menus and display the choice of the user in a sentence like: "You selected size 8 Nike shoes." If the user clicks "Other," an input statement should allow them to type a different shoe size (as a number). For this program, use `if-elseif` statements instead of `switch` statements.

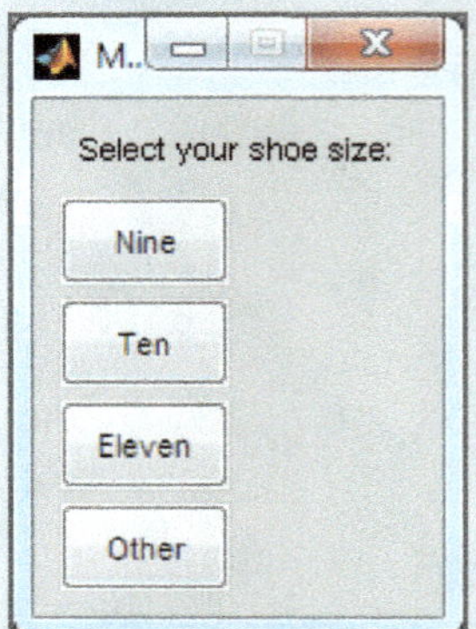

19. Assume you are required to generate the menus shown with Review Question 19-18 in a MATLAB program. Write the program necessary to generate these menus and display the choice of the user in a sentence like "You selected size 8 Nike shoes." If the user clicks "Other," an input statement should allow them to type a different shoe size (as a number). For this program, use `switch` statements instead of `if-elseif` statements.

20. The variable grade can have any real number values from 0 to 100. Ask the user to enter a grade in numerical form. Write an `if-elseif-else` statement that displays the letter grade (any format) corresponding to a numerical grade in an appropriately formatted output statement.

A typical range of grades:

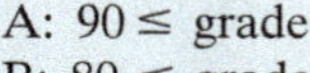

A: $90 \leq$ grade
B: $80 \leq$ grade < 90
C: $70 \leq$ grade < 80
D: $60 \leq$ grade < 70
F: grade < 60

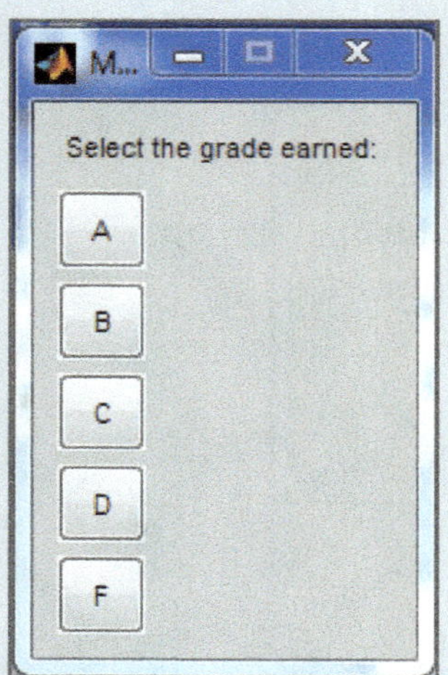

21. Write a MATLAB program that will allow the student to select the grade earned in a course from a menu (with the options of earning an A, B, C, D, or F) that will display the letter range for the selected grade. The output of the program displayed on the screen should be formatted as shown in the sample output. In your program, you must use a `switch` statement instead of an `if-elseif` statement. You may assume that the typical range of grades in a course is defined as listed in Review Question 19-20:

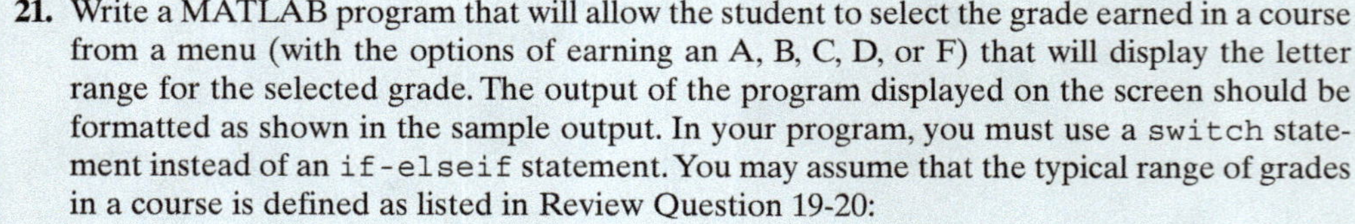

Sample Input / Output

```
if selected A...

If you earn a A, your earned grade is in the range: 90 <= grade.
```

22. *One of the fourteen NAE Grand Challenges is Engineering Better Medicines. Part of this Grand Challenge is creating individual medication plans for each patient, rather than sweeping recommendations based upon research using only one subset of the population. For example, in the past most heart attack protocols were created for white males. It has been discovered in recent years that women have different initial symptoms and respond better to a different treatment protocol than their male counterparts.*

To better tailor the medicine to the individual, you wish to write a computer program to allow the user to enter several variables, then the program will produce a suggested solution. Write the MATLAB code to . . .

(a) Ask the user to enter their name (text).
(b) Ask the user to type their weight (a number) in pounds-force.
(c) Ask the user to select their symptom from a menu, given the following choices.
- Cold
- Flu
- Migraine

(d) Based upon the answer selected to the menu, set the medicine name, volume [mL] and mass [g] of the medicine tablet.

Symptom	Medicine	Volume [mL]	Mass [g]
Cold	Achoo	3.5	9
Flu	Chill	5	16
Migraine	HAche	4	11

(e) Write (a) the function call from the program and (b) a function that will accept the mass and volume of the desired medicine from the main program and return the specific gravity of the desired medicine tablet to the program. Remember: The density of water is 1000 kilograms per cubic meter.
(f) Write (a) the function call from the program and (b) a function that will accept the weight of the person, specific gravity of the medicine, and tablet volume from the main program and return the number of tablets recommended for the person to the program. To determine the number of tablets needed, the following equation should be used and then rounded up to the next whole number of tablets. Any calculations or conversions necessary should appear in the function.

$$\text{Dose} = \text{Mass of person [kg]} \left| \frac{1.25 \frac{\text{mL}}{\text{kg}}}{2.5*\text{SG}} \right| \frac{1 \text{ tablet}}{\text{Volume[mL]}}$$

(g) Write an output statement that appears as follows, where `N` is replaced by the name of the medicine and `X.XXX` is replaced by the calculated specific gravity formatted to show 3 decimal places: `The specific gravity of N is X.XXX.`
(h) Write an output statement that appears as follows, where `AAAA` is the name of the user, `N` is the name of the medicine, `YYY` is the name of their symptom, and `Z` is the number of tablets recommended:

`AAAA, your recommended dosage of N to treat a YYY: Z tablets.`

NOTE: Although you may use conditional statements to solve this problem, none are needed if you set up the medicine data table as a cell array.

23. Assume two matrices, `M1` and `M2` have already been defined and have the same dimensions. Create a matrix `Comp1` with the same dimensions as `M1` and `M2`. Each element of `Comp1` should be an integer between −3 and 3 based on the relative values of the corresponding elements of `M1` and `M2` according to the following rules:

- If the magnitudes (absolute values) of corresponding elements of `M1` and `M2` are equal, that element of `Comp1` equals 0.
- If the magnitude of the element in `M1` is greater than that in `M2`, the corresponding element in `Comp1` will be positive.
- If the magnitude (absolute value) of the element in `M1` is less than that in `M2`, the corresponding element in `Comp1` will be negative.
- If the values in both `M1` and `M2` are positive, the magnitude of the value in `Comp1` is 1.
- If the values in both `M1` and `M2` are negative, the magnitude of the value in `Comp1` is 2.
- If the values in both `M1` and `M2` opposite signs (assume zero is positive), the magnitude of the value in `Comp1` is 3. (NOTE: if the magnitudes of the values in `M1` and `M2` are equal, that element of `Comp1` will contain a zero.)

Sample Input / Output

$$\texttt{M1} = \begin{bmatrix} 0 & 1 & 2 \\ -2 & -1 & 0 \\ 1 & -1 & -1 \end{bmatrix} \texttt{M2} = \begin{bmatrix} -1 & 1 & 0 \\ -1 & 2 & 2 \\ -1 & 0 & -2 \end{bmatrix} \texttt{Comp1} = \begin{bmatrix} -3 & 0 & 1 \\ 2 & -3 & -1 \\ 0 & 3 & -2 \end{bmatrix}$$

NOTE: Even if you know how to use programming loops, DO NOT do so for this program. The problem should be solved using logical arrays.

CHAPTER 20
LOOPING STRUCTURES

The key to making an algorithm compact is to exploit its variables by means of a loop. A loop is a programming structure that allows a segment of code to execute a fixed number of times or until a condition is true. This chapter describes two looping structures: the `for` loop and the `while` loop.

20.1 `for` LOOPS

LEARN TO: **Automate a segment of code using a `for` loop**
Calculate the number of executions of a `for` loop
Convert an algorithm into code involving a `for` loop

You may have noticed that a number of the algorithms we have already studied have parts that are repetitive. Sometimes, we want to execute an algorithm a certain number of times:

- For every pressure sensor, record the pressure reading.
- For each employee, check the date of the employee's last safety training.

It may be useful to sample less than every possible point, as in:

- For every fifth data point, enter its value into an array for plotting.

The ability to count backwards is also useful:

- For every second counting down from 10 to 0 seconds, announce the time remaining until launch.

The syntax of a `for` loop is illustrated below:

```
for counter = start : step : finish
        % executable statements
end
```

- ***counter:*** variable chosen to keep track of number of times the loop has executed
- ***start:*** first value of the loop
- ***step:*** incremental value to use to advance from start to finish
- ***finish:*** final value of the loop

The counter variable can be used in the loop to calculate other variables, to index an array, to set function parameters, etc.

NOTE

The variable *k* is defined in the header of the `for` loop. If you overwrite the value of *k* in the `for` loop, it might lead to unintended results.

A sample `for` loop is shown below. This loop will display on the screen the numbers from 1 to 5, each on a separate line.

```
for k=1:1:5
        fprintf('%0.0f\n',k)
end
```

Understanding the `for` Loop

Most program statements do exactly the same type of thing every time, although they may be manipulating different information. The `for` loop is slightly different. A `for` statement can be arrived at during program execution in one of two ways:

1. It can be executed immediately following a statement that was outside the loop, typically a statement immediately above it.
2. It can be executed upon returning from the `end` statement that marks the end of the loop after the loop has executed one or more times.

What the `for` statement does in these two cases is different and is crucial to understanding its operation.

1. When the `for` statement is arrived at from another statement *outside* the loop, it does the following:
   - Places the start value into the counter variable.
   - Decides whether or not to execute the instructions inside the `for` loop.
2. When the `for` statement is arrived at from its corresponding `end` statement below, it does the following:
   - Adds the step value to the counter variable.
   - Decides whether or not to execute the instructions inside the `for` loop.

The second step is the same in both cases and requires a bit more explanation. To determine whether or not to execute the instructions inside the loop (those between the `for` and its corresponding `end`), the `for` statement asks the question "Has the counter gone beyond the finish value?"

- If the answer is no, execute the statements inside the loop.
- If the answer is yes, skip over the loop and continue with the first statement (if any) immediately following the corresponding `end` statement.

NOTE

Since the loop can count either up or down, we use the word "beyond" to describe the comparison of the counter and the finish value.

If the loop is counting up (step > 0), then we could ask "Is the counter greater than the finish value?" If the loop is counting down (step < 0), we could use "less than" instead. Several examples are given below.

EXAMPLE 20-1

Write the command to execute the following loop the desired amount of times.

(a) The statements in the loop will be executed four times, once each with `i` = 1, 2, 3, and 4:

```
for i=1:1:4
        % loop statements
end
```

(b) The statements in the loop will be executed two times, once each with `time` = 1 and 3. When `time` is incremented to 5, it is "beyond" the finish value:

```
for time=1:2:4
        % loop statements
end
```

(c) The statements in the loop will be executed three times, once each with `gleep` = 4, 2, and 0:

```
for gleep=4:-2:0
        % loop statements
end
```

(d) The statements in the loop will be executed three times, once each with `raft` = 13, 7, and 1. When `raft` is decremented to −5 it is "beyond" the finish value:

```
for raft=13:-6:-4
        % loop statements
end
```

EXAMPLE 20-2

Determine how many times the loop will execute and for what values of the index variable.

(a)
```
for time=1:2:4
        % loop statements
end
```

The statements in the loop will be executed two times, once each with `time` *= 1 and 3. When* `time` *is incremented to 5, it is "beyond" the finish value:*

(b)
```
for raft=13:-6:-4
        % loop statements
end
```

The statements in the loop will be executed three times, once each with `raft` *= 13, 7, and 1. When* `raft` *is decremented to −5, it is "beyond" the finish value.*

(c)
```
for k=6:1:4
        % loop statements
end
```

The statements in the loop will not be executed. `k` *is "beyond" the finish value when the loop is first entered.*

(d)
```
S1=5;
S2=7;
for wolf=S1:S2:S1^2+S2+1
        % loop statements
end
```

The statements in the loop will be executed five times, once each with `wolf` *= 5, 12, 19, 26, and 33.*

As noted earlier, it is generally a bad idea to redefine the index variable inside of the loop. Similarly, redefining any variables from which the index variable is calculated is also strongly discouraged. For example, `S1` *and* `S2` *should not be changed inside the loop.*

Arithmetic Sequences

When dealing with **arithmetic sequences**—those that require only changing the **index variable** by adding or subtracting a number from it—there are two primary ways to proceed.

1. Set up the sequence explicitly, such as `k=2:2:10`
2. Set the index variable to count from 1 through the desired number of iterations, and then calculate the desired values from the index values, such as `k=1:1:5; k1=2k;`

In some cases the direct sequence may be more intuitive, in others the calculated equivalent makes more sense.

In each of the following examples, a sequence of five numbers is given, with two solutions shown to generate that sequence. The first code segment in each example uses the `for` loop to count from 1 to 5, and the values desired are calculated inside the loop. The second segment of code accomplishes the same goal by manipulating the `start`, `step`, and `finish` values instead. Although we are using print statements inside the example loops so that you can easily type these into MATLAB and see the sequence generated, you will seldom print the index values. Most of the time, the index values will be used in a calculation of some form, not just printed to the screen.

EXAMPLE 20-3

We desire the sequence to be 3, 6, 9, 12, 15.

```
for k=1:1:5
    fprintf('%0.0f\n',3*k)
end
```

Given that `k` *goes through the sequence k = 1, 2, 3, 4, 5, the desired sequence could be calculated as* `3k`.

```
for k=3:3:15
    fprintf('%0.0f\n',k)
end
```

This illustrates the idea that multiplication and division of a sequence by a constant to generate a new sequence affects all three loop control parameters: `start`, `step`, *and* `finish`.

EXAMPLE 20-4

We desire the sequence to be 3, 4, 5, 6, 7.

```
for k=1:1:5
    fprintf('%0.0f\n',k+2)
end
```

This can be calculated as `k` *+ 2*

```
for k=3:1:7
    fprintf('%0.0f\n',k)
end
```

Adding and subtracting a constant to create a new sequence affects only the `start` *and* `finish` *values.*

EXAMPLE 20-5

We desire the sequence to be 4, 6, 8, 10, 12.

Note that the standard order of operations applies, and the multiplication must be done first.

```
for k=1:1:5
    fprintf('%0.0f\n',2*k+2)
end
```

This can be calculated as $2k + 2$.

```
for k=4:2:12
    fprintf('%0.0f\n',k)
end
```

Note that all three loop parameters are affected: `start` *and* `finish` *are subject to both multiplication and addition, but* `step` *is only affected by multiplication.*

COMPREHENSION CHECK 20-1

Write a `for` loop to display every even number from 2 to 20 on the screen.

COMPREHENSION CHECK 20-2

Write a `for` loop to display every multiple of 5 from 5 to 50 on the screen.

COMPREHENSION CHECK 20-3

Write a `for` loop to display every odd number from 13 to −11 on the screen.

Using Variable Names to Clarify Loop Function

Using variables with meaningful names as loop controls makes it easier for those trying to interpret the program to keep track of what the loop is doing. The following loop keeps track of the position of an object moving at a constant speed. By defining the step distance as `Speed*TimeStep`, thus an increment in position, the number of steps in the loop is flexible as well. Note that these variables would need to be defined earlier in the program.

```
for Position=StartPosition:Speed*TimeStep:FinalPosition
     fprintf('%0.0f\n',Position)
end
```

Using a `for` Loop in Variable Recursion

An important use of loops is for **variable recursion**—passing information from one loop execution to the next. A simple form of recursion is to keep a running total. The loop below determines the total sales for all the vendors at a football game. The loop assumes that the array `Sales` contains the total sales for each vendor. The number of vendors is computed when the loop begins, using the `length` command, and the total sales are accumulated in `TotalSales`. Note that `TotalSales` must be initialized before the loop is entered, since it appears on the right-hand side of the equation:

```
TotalSales = 0;
for Vendor=1:1:length(Sales)
      TotalSales = TotalSales + Sales(Vendor);
end
```

This code also introduces a very important use of `for` loops: to automatically step through the values in a vector or array. Note here that the index variable `Vendor` is used sequentially to access each element of the `Sales` vector.

COMPREHENSION CHECK 20-4

Assume a vector `Vals` has already been defined. Write a `for` loop that will calculate the sum of all positive values in `Vals` and place the result in the variable `PosSum`.

Manipulating a `for` Loop Counter

Many kinds of computations can be performed in a loop. Many of the example loops above will output the value of `k` to the screen each time through the loop. Other sequences can be achieved by having an expression other than `k` in the executable part of the loop.

COMPREHENSION CHECK 20-5

Consider the following table of values. Determine the formulae to represent columns A through J. In some cases, the formulae in rows 1–10 will be similar. In other cases, the entry in the first row will be a number rather than a formula.

Create a `for` loop to display the values; use tabs in a formatted `fprintf` statement to create each column.

k	A	B	C	D	E	F	G	H	I	J
1	2	1	1	1	7.25	10	1	2	1	1.00
2	4	3	4	3	7.50	9	–1	4	2	0.50
3	6	5	9	6	7.75	8	1	8	6	0.33
4	8	7	16	10	8.00	7	–1	16	24	0.25
5	10	9	25	15	8.25	6	1	32	120	0.20
6	12	11	36	21	8.50	5	–1	64	720	0.17
7	14	13	49	28	8.75	4	1	128	5020	0.14
8	16	15	64	36	9.00	3	–1	256	40320	0.13
9	18	17	81	45	9.25	2	1	512	362880	0.11
10	20	19	100	55	9.50	1	–1	1024	3628800	0.10

Using the Counter Variable as an Array Index

The example above used the counter variable `Vendor` to access each element of the `Sales` array in sequence to add them together. If we wanted to add only the odd-numbered elements, the only change necessary would be to change the step variable to 2. If we wanted every fourth element of the array beginning with the fourth element, the `for` statement would become

```
for Vendor=4:4:length(Sales)
```

and so forth.

To access elements in a matrix, we could use a single index since MATLAB will handle this, but we would have to be very careful to make certain which row and column is being accessed. For example, if `A` is a 3×2 matrix, `A(4)=A(1,2)`. Since this can be quite confusing, it is usually better to use the double index notation when accessing values in a matrix. One method to accomplish this would be to use nested `for` loops, that is, a `for` loop inside another `for` loop. Let us modify the football sales example above as an example.

Assume we have a 2-D array named `Sales` where each row represents a specific vendor and each column represents a different category of item, such as drinks, hot dogs, hats, etc. Now if we want the total sales, we need to step through every element in every column (or every element in every row) and add them all up. To determine the number of vendors and items in the matrix, we use the `size` function, rather than the `length` function we used previously with the vector.

```
% Initialize the sum of all sales to 0
TotalSales = 0;
% Determine the number of vendors (rows) and items (columns)
[NumVendors, NumItems]=size(Sales)
for Vendor=1:1:NumVendors          % this will index the rows
      for Item=1:1:NumItems        % this will index the columns
            TotalSales = TotalSales + Sales(Vendor, Item);
      end
end
```

Note that the inner `for` loop (`Item` loop) goes through all items (columns) before exiting, at which point the outer `for` loop increments to the next vendor (row) and the inner loop resets to the first column and steps through all items again, but for a different vendor.

If there were four vendors (rows) and three items (columns), the order in which the entries in the `Sales` array would be added to `TotalSales` would be:

(1,1); (1,2); (1,3); (2,1); (2,2); (2,3); (3,1); (3,2); (3,3); (4,1); (4,2); (4,3);

You may recall from Chapter 16 that this problem could be solved more "simply" with a single line of code:

```
TotalSales=sum(sum(Sales));
```

There are many situations, however, where the built-in matrix operations will not accomplish the desired purpose and you MUST set up a pair of nested loops to step through the elements of the array.

COMPREHENSION CHECK 20-6

Write two nested `for` loops to determine how many positive values (including zeros) are in each column of the matrix `M1`, which has already been defined. The results should be stored in a row vector `PosNums` in which each element contains the number of positive values found in the corresponding column of `M1`.
Example: `M1=[3 -6 0;-4 -8 2;8 -9 1];` `PosNums=[2 0 3]`

Ending `for` Loops Early

In certain situations, it might be desirable to build in logic within a `for` loop that allows for the loop to terminate when a certain situation is encountered. The `break` command instructs MATLAB to exit the loop and continue as if the entire loop sequence was finished. It is important to realize that the break command does NOT terminate the function or program—it will only exit the loop code.

EXAMPLE 20-6

Consider the following segment of MATLAB code:

```
S = input('Type the starting point of the loop: ');
E = input('Type the ending point of the loop: ');
for i=S:1:E
     if 1/i == Inf
          fprintf('Uh oh! Division by zero!\n')
          break
     end
     fprintf('%0.1f\n',1/i)
end
fprintf('... ALL DONE!\n')
```

What is the outcome if the user enters S=1 and E=3?

When the user provides the value of 1 for the starting value and 3 for the ending value, everything runs smoothly:

```
Type the starting point of the loop: 1
Type the ending point of the loop: 3
1
0.5
0.3
... ALL DONE!
```

What is the outcome if the user enters S=-3 and E=3?

However, when the user provides the values of –3 for the starting value and 3 for the ending value, notice what happens:

```
Type the starting point of the loop: -3
Type the ending point of the loop: 3
-0.3
-0.5
-1
Uh oh! Division by zero!
... ALL DONE!
```

There are a few important concepts to note from the output of this execution of our code segment. First, the code after the `for` *loop, the output statement that displays "... ALL DONE!" is executed, so the* `break` *command will not skip any code after the* `for` *loop has terminated. Second, since the output of the number occurred after the check to see if a division by zero has occurred, the output command did not display the number zero because the* `break` *command terminated the loop, which means that no code after the* `break` *command within an iteration of a loop will execute. Last, it's important to note that the* `break` *command terminates the ENTIRE loop—not just the iteration, so the numbers 1 through 3 will not be displayed because the loop finished early.*

When a `break` command appears within a nested `for` loop, the `break` will apply to only the loop where the break occurs and will not terminate the entire nested loop.

EXAMPLE 20-7

Consider the following segment of MATLAB code:

```
Si = input('Type the starting point of the inner loop: ');
Ei = input('Type the ending point of the inner loop: ');
So = input('Type the starting point of the outer loop: ');
Eo = input('Type the ending point of the outer loop: ');

for i=So:1:Eo
    for j=Si:1:Ei
      if i == j
          fprintf('Same value... skipping!\n')
          break
      end
      fprintf('%0.0f\t%0.0f\n',i,j)
    end
end
fprintf('... ALL DONE!\n')
```

What is the outcome if the user enters 3, 5, 1, 2?

When the user provides the value of 3 for the starting value and 5 for the ending value of the inner loop and 1 for the starting value and 2 for the ending value of the outer loop, everything runs smoothly since these two sets of numbers will never intersect.

```
Type the starting point of the inner loop: 3
Type the ending point of the inner loop: 5
Type the starting point of the outer loop: 1
Type the ending point of the outer loop: 2
1    3
1    4
1    5
2    3
2    4
2    5
... ALL DONE!
```

What is the outcome if the user enters 3, 5, 3, 5?

However, when the user provides the values of 3 for the starting value and 5 for the ending value of both the inner and outer loops, notice what happens:

```
Type the starting point of the inner loop: 3
Type the ending point of the inner loop: 5
```

```
Type the starting point of the outer loop: 3
Type the ending point of the outer loop: 5
Same value... skipping!
4    3
Same value... skipping!
5    3
5    4
Same value... skipping!
... ALL DONE!
```

Notice that the outer loop is unaffected by the `break` *command and continues looping over values 3 to 5, regardless of whether the inner loop encounters a* `break` *command.*

EXAMPLE 20-8

Assume you're required to write a MATLAB program that will allow the user to generate a graph and plot a linear trendline of flowrate data manually provided by the user. Assume that your program will prompt the user to enter the initial time of data collection, the final time, and the interval between data samples, all in units of minutes. Your code should then iterate and prompt the user to record the corresponding flowrate sample value in units of gallons per minute at each required sample time.

```
clear
clc

AbsStart=input('What is the initial sample time [min]? ');
AbsEnd=input('What is the final sample time [min]? ');
AbsInc=input('How often should samples be taken [min]? ');

Abs = []; % create empty vectors to hold the user input
Ord = [];
cnt = 1; % cnt = count number of samples entered by user
for I=AbsStart:AbsInc:AbsEnd
    prompt=sprintf('Sample #%0.0f, at %0.2f min\nFlowrate (Q) [gpm]: ', cnt,I);
    Abs=[Abs I];
    x = input(prompt);
    Ord = [Ord x];
    cnt = cnt + 1;
end

% polyfit to calculate trend:
C = polyfit(Abs,Ord,1);
TrendAbs = [AbsStart:AbsInc/20:AbsEnd];
TrendOrd = C(1)*TrendAbs + C(2);

% generate plot
plot(Abs,Ord,'x',TrendAbs,TrendOrd,'-')
% other proper plot commands to show, for example, the axis labels, are
% entered here, but are not shown for space considerations
```

Sample Usage:

```
What is the initial sample time [min]?: 1
What is the final sample time [min]?: 6
```

```
How often should samples be taken [min]?: 2
Sample #1, at 1.00 min
Flowrate (Q) [gpm]: 7
Sample #3, at 3.00 min
Flowrate (Q) [gpm]: 17
Sample #5, at 5.00 min
Flowrate (Q) [gpm]: 27
```

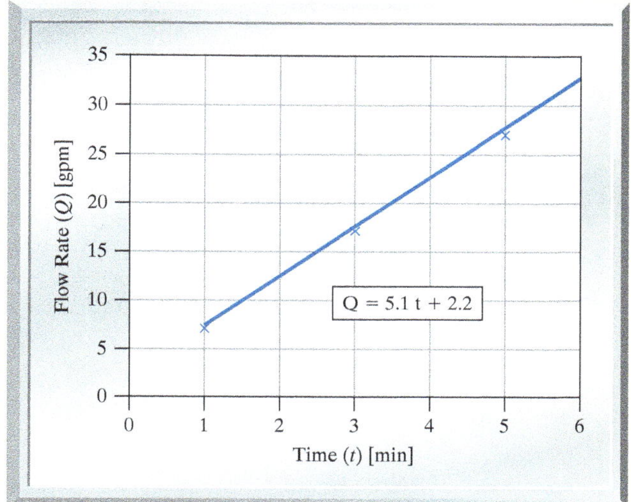

COMPREHENSION CHECK 20-7

Assume a matrix `M2` has already been defined. Write two nested `for` loops to determine how many rows of the matrix `M2` contain only negative values. The results should be stored in `NegRows`.
Example: `M2=[3 -6 0;-4 -8 -2;8 -9 1];`   `NegRows=1`

20.2 `while` LOOPS

LEARN TO: Automate a segment of code using a `while` loop
Calculate the number of executions of a `while` loop
Convert an algorithm into code involving a `while` loop

Once a `for` loop has started, it executes a specified number of times, regardless of what happens in the loop. At other times, we want to continue executing a loop until a particular condition is satisfied. This requires the use of a `while` loop, which executes until the specified condition is false. The conditional part of the `while` statement has the same syntax as that of the `if` statement.

```
while Logical_Expression
        % executable statements
end
```

For example, while the calculation error is unacceptable, refine the calculations to improve accuracy.

```
while error >= 0.01
        % require less than 1% error
        % perform another iteration of the calculation
        % recalculate error
end
```

Converting "Until" Logic for Use in a `while` Statement

In some cases, it may make more sense to use the word *until* in phrasing a conditional loop, but MATLAB does not have such a structure. As a result, it is sometimes necessary to rephrase our conditions to fit the `while` structure.

"Until" Logic Condition	`while` Logic Translation

`until cows == home`
`party`
`end`
	while cows ~= home
	party
	end

`until homework == done`
`TVpower = off`
`end`
	while homework ~= done
	TVpower = off
	end

Initializing `while` Loop Conditions

In `for` loop constructions, the loop initializes and keeps track of the loop counter within the loop command. In a `while` loop, it is necessary to make sure that variables are properly initialized before entering encountering the loop command.

```
error=1;
while error >= 0.01
        % require less than 1% error
end
```

If `error` is not initialized, a syntax error results since the `error` term does not exist when MATLAB tries to compare it to the value 0.01. If `error` is incorrectly initialized to a value less than 0.01, the loop never executes. The initial value is chosen at random; we could have made the value any number greater than 0.01 and the program would execute correctly.

EXAMPLE 20-9

We want to input a number from the user that is between (and including) 5 and 10, but reject all other numbers. All extraneous output should be suppressed.

```
X=0;
while X < 5 || X > 10
    X=input('Enter a number between 5 and 10');
end
```

COMPREHENSION CHECK 20-8

Write a `while` loop that requires the user to input a number until a nonnegative number is entered.

EXAMPLE 20-10

We have a number stored in a variable `T` that we want to repeatedly divide by 10 until `T` is smaller than 3×10^{-6}. After determining the first value of `T` that meets the condition, it should be displayed in a formatted `fprintf` statement where the value of `T` is displayed in exponential notation. All extraneous output should be suppressed.

```
T=3993;
while T > 3E-6
        T=T/10;
end
fprintf('%E\n',T);
```

EXAMPLE 20-11

Given a vector of positive integers, `V`, we want to create two new vectors: vector `Even` will contain all of the even values of `V`; vector `Odd` will contain all of the odd values of `V`. To begin solving this problem, we need to seek out a MATLAB function that will help us to determine whether or not a given value is even or odd. The built-in function `rem` accepts two input arguments (`Z = rem(X,Y)`) such that `Z` is the remainder after dividing `X` by `Y`. For example, if we type `rem(3,2)`, the function would return `1` because $3/2 = 1$ with a remainder of 1.

```
V=[2, 3, 91, 87, 5, 8];
Even=[];
Odd=[]; % initially, Even and Odd are both empty vectors
fprintf('\nV =\t%0.0f', V')
while length(V)>0
        if rem(V(1),2)==0       % is first element of V even?
            Even=[Even V(1)]; % place it in Even
        else                    % is first element of V odd?
            Odd=[Odd V(1)];     % place it in Odd
        end
        V(1)=[];                % delete the first element of V
end
fprintf('\nEven =\t%0.0f', Even')
fprintf('\nOdd =\t%0.0f', Odd')
fprintf('\n')
```

The output of this code segment would be:

```
V    =    2    3    91    87    5    8
E    =    2    8
O    =    3    91   87    5
```

COMPREHENSION CHECK 20-9

Assume a vector `V2` has already been defined and contains only values greater than 1. Write a `while` loop to calculate the product of the elements of `V2` in sequence from the first element until the product is greater than 10^6 or all elements have been included in the product. The result should be stored in `Prod`.

Example: `V2=[2 10 7 19];` `Prod=2,660`

Example: `V2=[3 6 123 4 58 267 8 91 11];` `Prod=137,144,016`

Custom `while` Loop Termination

Just like the `for` loop, the `while` loop allows code to be written in such a way that the `break` command will terminate the loop early. Note that the behavior of the `break` command is the same with a `while` loop— it will only end the loop early and will NOT terminate a program or function. After encountering the `break` command, the code continues executing at the next line after the end of the loop.

Consider the following `while` loop, where the loop will execute while "true." This is a common approach to designing algorithms where the looping logic is indeterminate or too complicated to efficiently implement with a pure `while` loop or `for` loop structure.

```
S = 0;
while true
      A = input('Type the alpha value: ');
      if A == 5
        break
      end
      S = A + S;
      B = input('Type the beta value: ');
      if B == 7
        break
      end
      S = 2*B + S;
      end
fprintf('The final answer is: %0.0f\n',S)
```

Consider the following output sequence from the code segment shown.

```
Type the alpha value: 1
Type the beta value: 2
Type the alpha value: 4
Type the beta value: 5
Type the alpha value: 5
The final answer is: 19
```

Note the loop terminates before the alpha value of 5 can be added to the summation variable. Likewise, if the user never types the values that trigger the `break` commands (alpha value of 5 or beta value of 7), this code will loop "infinitely". In practice, this code will eventually terminate once the value stored in the variable is "overflown" with a value greater than the variable type can handle, but if the user continues manually typing in small numbers, this will not happen for a long time.

Recall to end an indefinite loop use CTRL+C

20.3 APPLICATION OF LOOPS: GUI

LEARN TO: **Create graphical user interfaces (GUI) in MATLAB**
Incorporate the appropriate user interface controls on GUIs

The purpose of this section is to familiarize you with creating **graphical user interfaces** (**GUIs**—often pronounced "gooey") in MATLAB and provide some insight into user-centered design to give you the tools you need to build a robust but clear interaction platform without the much "heavy" work on behalf of the program user. Designing user interfaces is different from designing algorithms to implement as programs or functions because there is an additional layer of complexity necessary to obscure some of the programming or language-specific requirements of your code. Graphical interface design is especially important when delivering a final "usable" product to a client or customer (the user) who may not necessarily have knowledge of the programming language you used to implement your solution.

Different graphical interface systems have different sets of best practices. For example, Microsoft Windows applications tend to have boxes in the upper right corner that allow users to click buttons to minimize, maximize, or close a program, whereas Mac OS X applications contain three circles in the upper left corner of windows that perform similar functions. Graphical interfaces also exist in many mobile operating systems like iOS or Android that allow handheld devices like iPhones and iPads to interact using push buttons or touch gestures to complete tasks. The remainder of this document focuses on WIMP interaction—"window, icon, menu, pointing device" like that found on programs run in Microsoft Windows (like MATLAB itself—see Fig. 20-1) or Mac OS X, but similar controls and development ideologies exist for mobile applications.

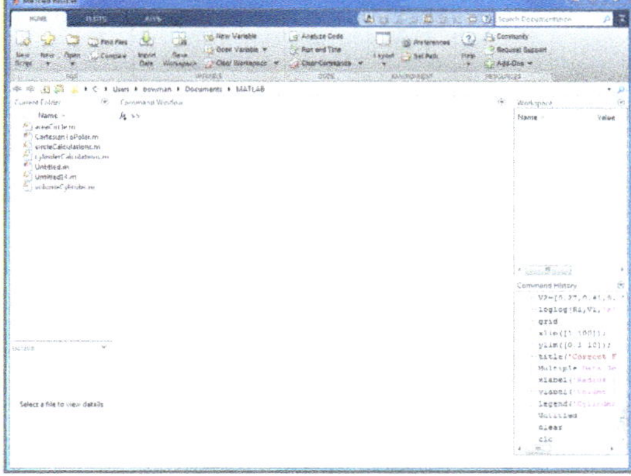

Figure 20-1 MATLAB itself is built with a GUI that allows you to click different toolbar buttons, menu items, launch new windows like the Editor Window, or type commands using the keyboard into the Command Window.

Graphical User Interface Development Environment

In MATLAB, the tool used for designing a user interface is the Graphical User Interface Development Environment (GUIDE). GUIDE can be launched in the main MATLAB window by selecting New > Graphical User Interface from the Home tab or by typing **guide** into the Command Window. The GUIDE Quick Start window (Fig. 20-2) gives you the options to either create a new GUI or open an existing GUI created with GUIDE. By default, there are a few built-in templates that give examples of how interactions are built using MATLAB commands with the GUI, but in general, most GUIs in MATLAB are built by selecting Blank GUI, the default option, and clicking the OK button.

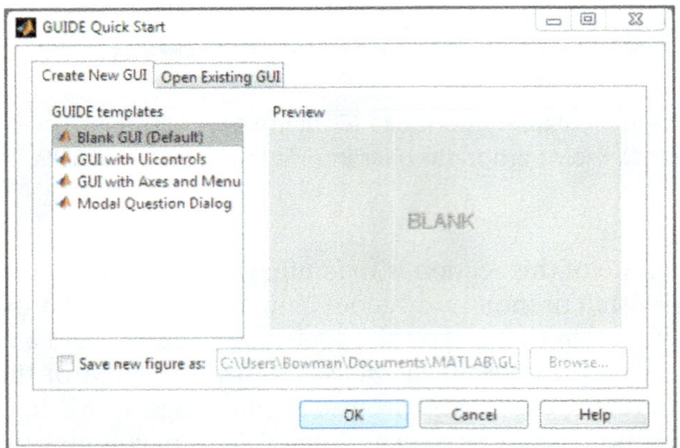

Figure 20-2 GUIDE Quick Start window.

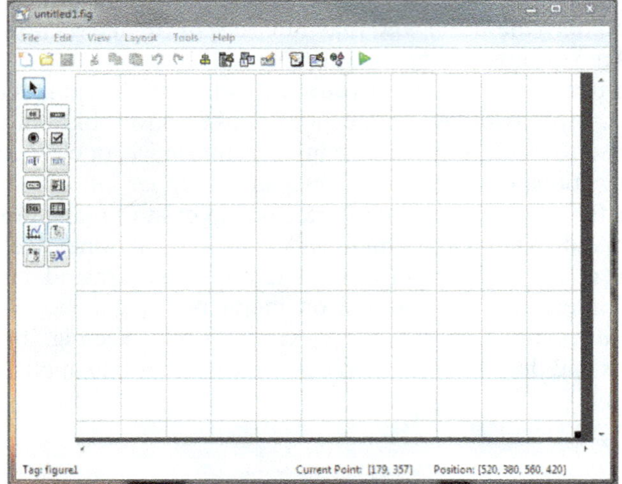

Figure 20-3 Graphical layout editor in GUIDE.

The resulting window (Fig. 20-3) is the graphical layout editor where you can arrange different user interaction controls that will allow, for example, the user of your program to click buttons, select options from check boxes or drop down menus, type values into text boxes, or an assortment of other options. This graphical layout is saved as a MATLAB Figure file with the file extension ".fig." For any GUI built in MATLAB using GUIDE, there will be both an M-File and a Figure file—both named the same and following the standard set of MATLAB file naming rules.

The graphical layout editor contains a set of buttons on the left side of the window that allows you to insert different interface controls on your GUI. To the right of the button set, the canvas of the GUI allows you to lay out exactly where the different interface controls should appear on the window of your application. Finally, the status bar at the bottom of the window contains different information depending on which user interface control is selected on the canvas. By default, the "figure" control is selected with the tag name "figure1"—this will be the main area where other controls will be placed in each window. In addition, as you move your mouse over the canvas, the "Current Point" value in the status bar updates with the current XY coordinate of your mouse within the GUI. That position will be helpful in lining up interface controls on your GUI so that it looks clean and consistent. In Table 20-1, the different user interface controls are described along with typical examples of when these controls might be used in a program window.

Table 20-1 User interface controls in MATLAB

Control	Name/Description	Sample Uses	Not Used For . . .
OK	**Push Button** Launches a calculation/action or opens a new dialog window.	Submit/calculate action, quit program, close window, load/open data file, cancel operation.	Setting or unsetting a parameter—use a checkbox or toggle button instead.
	Slider Slides a value between a minimum and maximum value by a step size.	Increasing or decreasing a variable value, incrementing or iterating through an expression.	Setting a value outside of the minimum or maximum value of the range—use a text box instead (or in addition to the slider).
	Radio Button Allows the user to select a single value from a mutually exclusive list of options. These are generally used inside of button group controls.	Selecting gender (male or female), water phase (solid, liquid, or gas), or other options from small groups.	Groups that have more than 5 or 6 different options (unit systems, when all SI units are included)—use a pop-up menu instead.
	Check Box Allows the user to turn on (or off) a single value. In general, check boxes are presented as a group of check boxes.	Selecting font style: bold, italic, underline. The user may want to use 1, 2, or all 3 options, or select none.	Allowing conflicting selections—a person can't be both male and female! Use a radio button instead.
EDIT	**Edit Text** Allows the user to manually type data into the program. Also called "entry fields" in some GUI environments.	Typing text (first name, last name, phone number) or numbers (zip code, weight, height, tire pressure).	Typing an element from a closed set of options—you don't want the user to type "Male"—let them select it using a radio button.
TXT	**Static Text** Allows the program to display or update text in a control that is not editable by the user. Also called "labels" or "protected fields" in some GUI environments.	Display program header labels, text entry field labels, slider values, calculated values, small error messages, or other information of interest to the program user.	Editing or typing text—use an edit text control instead.
	Pop-up Menu Allows the user to select a single option from a list instead of typing values in an edit text box.	Selecting state of birth from a list of 50 states, selecting a car manufacturer from a list, selecting the car model from a list.	Pop-up menus don't allow the user to select more than one option (e.g. Jeep and Dodge)—use a listbox control instead.
	Listbox Allows the user to select multiple options from a list instead of typing multiple values into multiple text boxes.	Select multiple beam lengths, pipe sizes, data sets for a plot, students in an engineering class, or other scenarios where you want to choose more than one option.	If you want to restrict the selection to only one option, use a pop-up menu instead.

(Continued)

Table 20-1 (Continued)

	**Toggle Button** Allows the user to turn on (or off) a value. Toggle buttons can be presented in groups, but can also stand alone. For exclusive selection of toggle buttons, use the toggle controls inside a button group.	Turning on or off debug output. Selecting font style: bold, italic, underline. The user may want to use 1, 2, or all 3 options, or select none. Can be used like a check box.	Can only be used in binary states (on or off)—for more than 2 states, use a control like a radio button or pop-up menu.
	**Table** Allows the program to display tabular data in a visually attractive and interactive (e.g. scrollable) control.	Displaying a list of spring stiffnesses for multiple springs. Displaying a matrix of values calculated by a program.	Not useful for displaying single text strings or values—use a static text control instead.
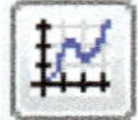	**Axes** Allows the program to display a plot or other data that can be displayed on axes in MATLAB (like an image) on the program window.	Displaying a plot of data, displaying an image of a card from a deck of cards, displaying the frequency response of an audio signal.	For multiple graphs on a GUI, you might want to consider using a single axis control and a subplot to allow MATLAB to automatically handle graph alignment.
	**Panel** Allows the programmer to group similar user interface controls to make the window visually easier to navigate.	Grouping all of the input controls in an "input" panel, grouping all of the output controls in an "output" panel.	Does not enable exclusivity within controls like radio or toggle buttons—use a button group if you require exclusive selection of options.
	**Button Group** Allows the programmer to group radio and/or toggle buttons to allow the user to make a mutually exclusive (single) selection from a group of multiple controls.	Selecting gender (male or female), water phase (solid, liquid, or gas), or other options from small groups.	Groups that have more than 5 or 6 different options (unit systems, when all SI units are included)—use a pop-up menu instead.
	**ActiveX Control** Allows the programmer to embed a control from a different program installed on the computer.	Embedding an Adobe PDF document in a GUI, embedding a QuickTime movie player.	Cross-platform programs—many ActiveX controls are operating system specific, so a GUI built-in Windows may not run in Mac OS X or Linux versions of MATLAB.

Interacting with User Interface Controls in GUIDE

Before laying out a program interface using GUIDE, it is critical to understand a few key ideas with user interface controls—specifically, how to change or extract values or properties of controls that are set automatically by your program. To demonstrate this, if you add a push button to the canvas in GUIDE, the tag (see the bottom left of the status bar in GUIDE when the push button is selected) of the push button is

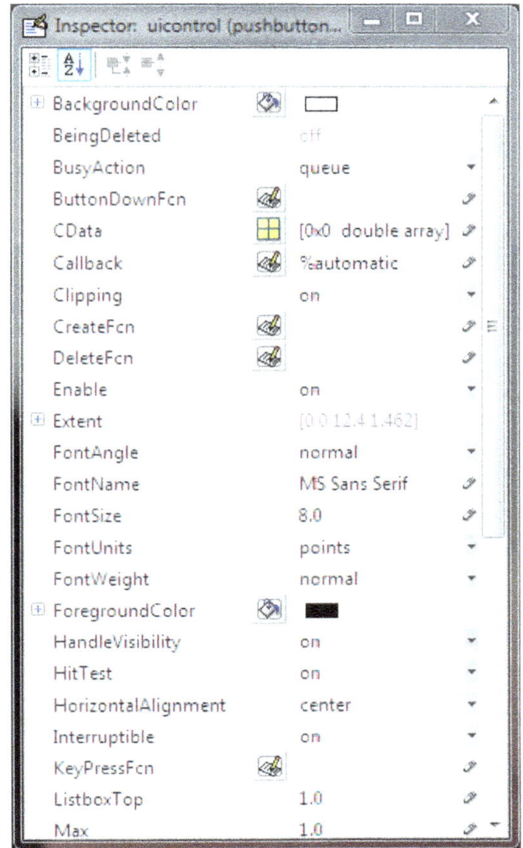

Figure 20-4 Property Inspector window for a push button control.

automatically set as "pushbutton1" by GUIDE. With this control, and all other controls, "tag" is a user interface control property that uniquely identifies that specific control on a GUI, much like how MATLAB would not allow you to have multiple variables with the same name. For each user interface control, there are a number of different properties that are automatically set by GUIDE, and understanding how those properties behave will allow for greater customization of the user interface.

The Property Inspector window allows the programmer to manually change the values in GUIDE of different user interface controls. In Fig. 20-4, the Property Inspector window shows the different properties for a push button user interface control that can be changed by typing different values if the property box contains a pencil icon or selecting different values on properties that contain a pop-up menu of different property values. For example, in this window, several properties like `FontName`, `FontSize`, `FontAngle`, and `FontWeight` will allow for customization of the font on the push button. Setting or getting the values of GUI controls can also be done in code using the `get` and `set` functions.

Interacting With User Interface Controls Programmatically: `get` and `set`

Before discussing the use of the `get` and `set` function to change or store property values in user interface controls, recall the use of the built-in graphing tool `fplot`, which we use for displaying functions in a MATLAB Figure window. In this code segment, we are plotting the `sin` function from $-\pi$ to π:

```
fplot('sin(x)',[-pi,pi]);
xlabel('X-Axis Text','FontSize',12,'FontWeight','bold');
ylabel('Y-Axis Text','FontSize',12,'FontWeight','bold');
grid
```

In this code segment, we are using the extra arguments in the `xlabel`, `ylabel`, functions to change properties like `FontSize`, `FontWeight`, by listing the property names and values as extra arguments in the function call. All plotting functions like `fplot` or `plot` actually create GUIs that are built in to MATLAB that have specialized functions like `xlabel`, `ylabel`, that are simply just short-cut tools that manually set the text values of the title and axis labels on plots. Since these plot windows are GUIs, the title and axis labels are actually just user interface controls like the static text labels described in Table 20-1, so these controls can actually be manually set or referenced creating a handle to the user interface control. A handle is actually just a reference that knows which control we want to modify that will allow us to access the different properties of the user interface.

To set user interface control properties using the `set` function, we need to pass in three arguments:

```
set(h,'PropertyName','PropertyValue');
```

where h is the variable that contains the handle reference, 'PropertyName' is the name of the property (like 'FontWeight' in the code above), and 'PropertyValue' is the value to set (like 'bold' in the code segment above).

The equivalent code to the code segment above written using set function calls:

```
fplot('sin(x)',[-pi,pi]);
xl=xlabel('X-Axis Text');
yl=ylabel('Y-Axis Text');
t=title('Title Text');
set(xl,'FontSize',12);
set(yl,'FontSize',12);
set(xl,'FontWeight','bold');
set(yl,'FontWeight','bold');
grid
```

Note that in the block of MATLAB code above, the handles to the user interface controls are created by capturing the handle references in different variables (xl, yl,) as outputs of the xlabel, ylabel, functions. Plotting functions like xlabel and ylabel were designed to be able to modify property values using extra function input arguments to reduce the number of lines of code when writing common code like creating graphs, so that is the preferred method for setting properties with those functions. The use of the set function in this fashion is only presented here as an example to help explain how this function works later when presented in the context of GUIs.

The get function works similarly to the set function by referencing user interface controls using a handle:

```
V=get(h,'PropertyName')
```

where h is the handle reference to the user interface control, 'PropertyName' is the property in the control, and V is a variable that will contain the property value for 'PropertyName'.

Consider the following segment of MATLAB code that uses the fplot function to again plot sin(x) from $-\pi$ to π and displays a title on the graph.

```
fplot('sin(x)',[-pi,pi]);
t=title('Title Text');
fs=get(t,'FontSize');
fw=get(t,'FontWeight');
fc=get(t,'Color');
st=get(t,'String');
```

In this code segment, we use the get function to read the default property values for the FontSize, FontWeight, Color, and String property values into the variables fs, fw, fc, and st. Note that fw and st are both text variables, fs is a number, and fc is a vector that contains the amount of red, green, and blue in the text font face color.

EXAMPLE 20-12

In this example, the canvas (Fig. 20-5) contains a simple GUI with a few static and edit text boxes and a push button that will allow us to multiply the numbers typed into the boxes on the GUI and display them in the interface.

To begin, the controls were laid out on the GUI canvas as shown below. A list of the properties changes for the different user interface controls are documented below as well.

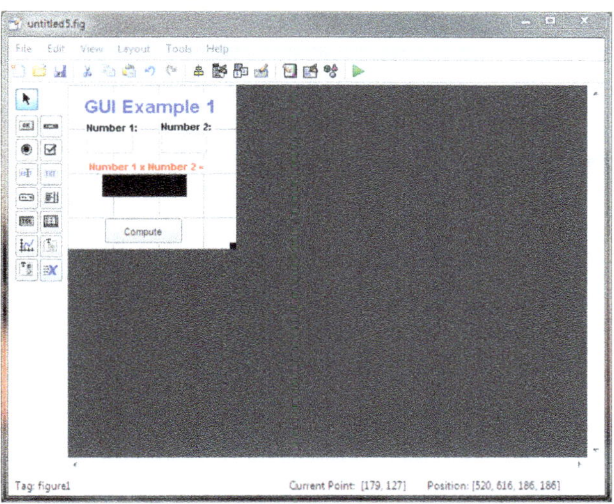

Figure 20-5 GUI for Example 20-12.

Tag: labelGUIHeader (static text)
String: GUI Example 1
ForegroundColor: blue
FontSize: 16
FontWeight: bold

Tag: labelNumber1 (static text)
FontWeight: bold
String: Number 1:

Tag: labelNumber2 (static text)
FontWeight: bold
String: Number 2:

Tag: label Multiplication (static text)
FontWeight: bold
String: Number 1 × Number 2
ForegroundColor: red

Tag: num2 (edit text)
String: (blank)

Tag: labelResult (static text)
FontSize: 14
FontWeight: bold
ForegroundColor: white
BackgroundColor: black
String: (blank)

Tag: num1 (edit text)
String: (blank)

Tag: button Compute (push button)
String: Compute

In addition to the changes above, the canvas was resized so that the GUI window does not contain a bunch of unnecessary unutilized space. After the GUI is properly arranged and renamed on the canvas, click the "Save and Run" button on GUIDE and the dialog shown in ▶.

Figure 20-6 will pop-up on the screen. Click Yes on the question dialog and type the name for the GUI Figure .fig file: GUIExample1.

Next, two windows will pop-up—the MATLAB Figure (Fig. 20-7) generated by GUIDE, as well as an M-File (Fig. 20-8) containing over 100 lines of MATLAB code and comments necessary to interact with the GUI. All of the functions within the M-File are actually

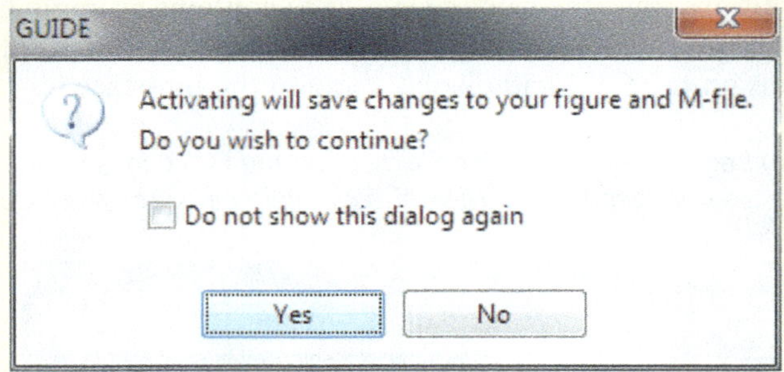

Figure 20-6 GUIDE save changes dialog.

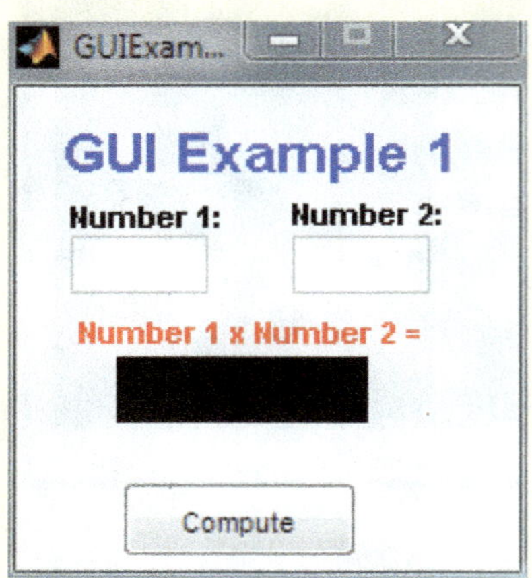

Figure 20-7 MATLAB Figure generated by GUIDE in Example 20-12.

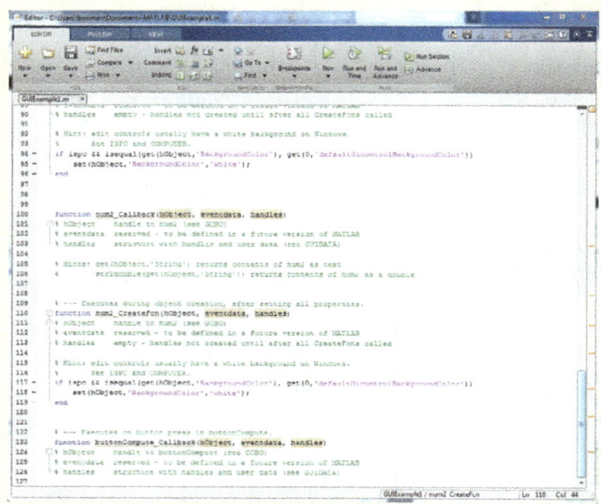

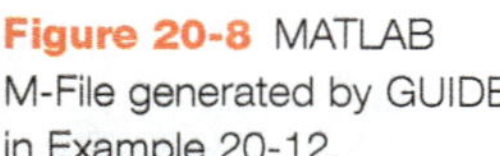

Figure 20-8 MATLAB M-File generated by GUIDE in Example 20-12.

subfunctions within the M-File, which means they can be called anywhere within this M-File, but you cannot call any of these functions from a different M-File. Subfunctions can only be written in M-Files that are implemented as a function, so the main GUI code is actually written as a function. Note that the M-File is named GUIExample1.m, which is the same file name as the MATLAB Figure we typed earlier. It is important that you do not rename this file since the Figure and M-File contain special references to "GUIExample1" throughout the code.

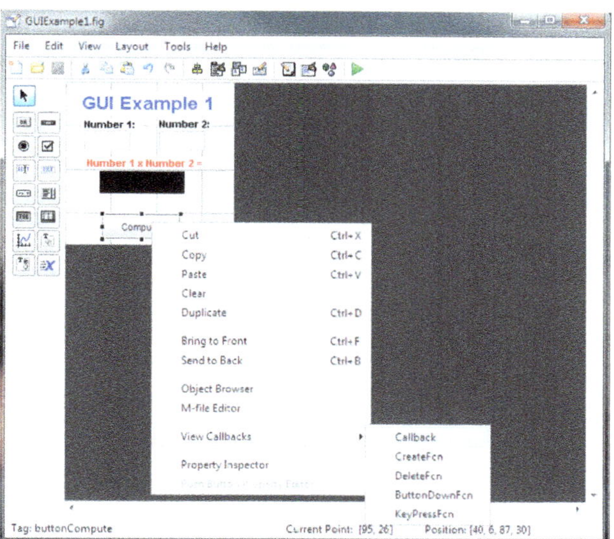

Figure 20-9 Accessing the push button callback function.

Now that the MATLAB Figure and M-File for the GUI exist, edit the M-File to insert some of the back-end code to implement the multiplication. To do this, we need to modify the callback function for the push button. Callback functions are codes that run as soon as some interaction occurs on the user interface. Examples of interaction might include clicking, double clicking, selecting an option from a menu, clicking a check box, or even events as simple as moving the mouse over a user interface control. To edit the callback function (as seen in Fig. 20-9) for the push button, right click the Compute button in GUIDE and select View Callbacks > Callback from the context menu.

At this point, the M-File Editor should pop-up and display the automatically generated code that handles the event that fires when the user clicks on the Compute button in your GUI:

```
% --- Executes on button press in buttonCompute.
function buttonCompute_Callback(hObject, eventdata, handles)
% hObject handle to buttonCompute (see GCBO)
% eventdata reserved - to be defined in a future version of
% MATLAB
% handles structure with handles and user data (see GUIDATA)
```

From here, we need to use the `get` *and* `set` *functions to pull down the values in the edit text boxes, multiply the values together, and display the result on the GUI. To do this, the function header generated by MATLAB contains a function input called handles that is a structure variable that contains handle references to all of the user interface controls in the GUI. We can access the different user interface controls by typing* `handles.tagName`*, where* `tagName` *is the actual tag name embedded in different control properties. In this example, if we wanted to access the contents of the* `num1` *edit text box, we could type the following:*

```
Num1Text=get(handles.num1,'String');
```

When we access some of the properties from user interface controls, be aware that some of the variable types might not be what you expect. In this case, the `Num1Text` *variable is actually a string of text instead of a number. Before we do our computation, we must take the strings we* `get` *from the two edit text boxes and convert them into numbers before we can multiply them. Likewise, the* `String` *property in the static text field that will contain the computed result expects the property to be set as a string, so we must convert the computed value into a string before we use the* `set` *function with the* `String` *property. The final code in the* `buttonCompute` *callback function should look something like:*

```
% --- Executes on button press in buttonCompute.
function buttonCompute_Callback(hObject, eventdata, handles)
% hObject handle to buttonCompute (see GCBO)
% eventdata reserved - to be defined in a future version of
% MATLAB
% handles structure with handles and user data (see GUIDATA)
Num1Text=get(handles.num1,'String');
Num2Text=get(handles.num2,'String');
M=str2double(Num1Text) * str2double(Num2Text);
set(handles.labelResult,'String',num2str(M,'%.0f'));
```

Callback and Related Functions for Different Controls

Like we saw in Example 20-12, many of the different user interface controls use callback functions or similar functions that will run when the user clicks on a control or somehow interacts with your GUI. These callback functions will only run the code within the function when the control is activated and will run again the next time the control is clicked. This idea is referred to as event-driven programming, where code will only execute when the user interacts with a control in an interface. User interaction with a control is referred to as an event, and when a callback function is executed, it is often said that the event for that control has fired.

In Table 20-2, the list of the important functions on different user interface controls is given with important properties that contain information about the text, values, or other data that might be necessary to use in the callback function. Note that there are other callback functions that may not be listed in the table that will fire on certain events like pressing a certain key on the keyboard, but those are not explicitly mentioned because they behave similarly to the standard callback functions.

Table 20-2 User interface controls

Control	Callback	Important Properties	Fires When . . .
Push Button	`Callback`	**'String'** – text on the button	Button is pressed
Slider	`Callback`	**'Value'** – current value of slider, number **'Min'** – minimum slider value, number **'Max'** – maximum slider value, number	Slider value changes
Radio Button	`Callback`	**'Value'** – current value of radio, number **'Min'** – minimum radio value (not selected), number **'Max'** – maximum radio value (is selected), number	Radio button is pressed

(Continued)

Table 20-2 (Continued)

Control	Callback	Important Properties	Fires When . . .
Check Box	`Callback`	**'Value'** – current value of radio, number **'Min'** – minimum radio value (not selected), number **'Max'** – maximum radio value (is selected), number	Check box is clicked (checked or unchecked)
Edit Text	`Callback`	**'String'** – text in the edit text box	After editing text, user clicks outside of the textbox
Static Text	`N/A`	**'String'** – text in the static text box	N/A
Pop-up Menu	`Callback`	**'String'** – the pop-up menu contents, as a cell array of text values **'Value'** – the index into the cell array of the selected item in the pop-up menu	Selection in the pop-up menu is changed
Listbox	`Callback`	**'String'** – the listbox contents, as a cell array of text values **'Value'** – the index or indices into the cell array of the selected item in the pop-up menu, either a number or a vector depending on the number of selected items	Selection in the listbox is changed
Toggle Button	`Callback`	**'Value'** – current value of the toggle button, number **'Min'** – minimum toggle button value (not pressed), number **'Max'** – maximum toggle button value (pressed), number	The user toggles the toggle button
Table	`N/A`	**'Data'** – the numerical values in the table **'ColumnName'** – the names of the columns, as a cell array **'RowName'** – the names of the rows, as a cell array	N/A
Axes	`N/A`	The standard plot functions will work with the axes control. If the axes handle name is axes1, the plot commands need to be preceded by `axes(handles.axes1)`	N/A
Panel	`N/A`	The panel control is generally used for layout purposes on the GUI, but does contain some functions that will fire if the GUI is resized	N/A
Button Group	`Selection-ChangeFcn`	The tag name of the selected button can be extracted by calling `get(eventdata.NewValue,'Tag')`	The user changes the selected object in the button group

Built-In Dialogs and Interaction Interfaces

There are a number of other built-in dialog and interaction interfaces that can complement your GUI by allowing all interaction to appear in the user interface rather than occasionally displaying or typing values in the Command Window. These functions include:

Built-In Dialog Boxes

- `dialog`: Displays text in a clickable textbox as a pop-up dialog.
- `msgbox`: Displays text in a clickable textbox as a pop-up dialog with customizable icons and dialog title.
- `errordlg`: Displays text in a clickable textbox as a pop-up dialog as an error message.
- `helpdlg`: Displays text in a clickable textbox as a pop-up dialog as a help message.
- `inputdlg`: Prompts the user to type in a value into an input dialog.
- `listdlg`: Prompts the user to select an option from a listbox in a pop-up dialog.
- `questdlg`: Prompts the user to answer a question (yes/no) in a question dialog.
- `warndlg`: Displays text in a clickable textbox as a pop-up dialog as a warning message.
- `printdlg`: Displays the print dialog to allow the user to print your GUI.

Built-In Interaction Dialogs

- `uigetdir`: Allows the user to select a directory that contains files.
- `uigetfile`: Allows the user to select a file (e.g., select an Excel file to open with `xlsread`).
- `uiputfile`: Allows the user to save a file (e.g., select where and what to name an Excel file you are writing with `xlswrite`).
- `uisetcolor`: Allows the user to pick a color from a color list.
- `uisetfont`: Allows the user to change font properties.

For more information on any of these functions, refer to the MATLAB `help` and `doc` pages for each function.

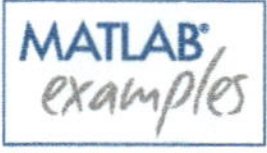

For more examples of GUI, look at the following .zip files available in the online materials:

- SliderAndPlotExample
- UIGetFileAndPlotExample
- UITableExample

In-Class Activities

ICA 20-1

For each code segment below, determine the contents of the specified variable following execution. If an error will occur, write ERROR and explain what caused the error.

(a) What is stored in Count by the following code?

```
Count=0;
for N=10:-0.2:8.5
    Count=Count+1;
end
```

(b) What is displayed on the screen by the following code?

```
for m=3:8
    fprintf('%0.0f\n',m^2-m)
end
```

(c) What is displayed on the screen by the following code?

```
for m=3:8
    M2=m^2-m;
end
fprintf ('%0.0f\n',M2)
```

(d) What is stored in P and C?

```
S=3;
I=2;
P=1;
C=0;
for K=S:I:S^I
    P=P*K;
    C=C+1;
end
```

ICA 20-2

Write a function named `CountDown` that accepts a total time `T` in seconds and displays the time remaining after half of the previously displayed time has elapsed as described below. The function should also accept an integer `Steps` as also described below.

The function should immediately display the value of `T`, and then wait `T/2` seconds, before displaying `T/2`. It should then wait `T/4` seconds and display `T/4`, etc. This should continue until only $1/2^{Steps}$ of the original time remains, at which point it waits for this remaining time, displays a zero, and emits an audio tone.

It is suggested that you look up the `pause` function and the `beep` function.

Example: `CountDown(60,6)` displays the following followed by an audio tone.

```
60          Displayed immediately
30          Displayed 30 seconds later
15          Displayed 15 seconds later
7.5000      Displayed 7.5 seconds later
3.7500      etc.
1.8750      etc.
0.9375      etc.
0           Displayed 0.9375 seconds later accompanied by a beep
```

ICA 20-3

For each code segment below, determine the contents of the specified variable following execution. If an error will occur, write ERROR and explain what caused the error.

(a) What is stored in VC?

```
VC=0;
V1=[9 5 -3 6 -1 0];
for Val=1:length(V1)
   if  V1(Val) <=0
      VC=VC+1;
   end
end
```

(b) What is stored in VS?

```
V2=[9 5 -3 6 -1 0];
for Val=2:length(V2)
   VS(Val-1)=V2(Val-1)+V2(Val);
end
```

(c) What is stored in VE?

```
VE=0;
V1=[9 -4 -3 3 6 -1 0 5 5 -2 -2 -3];
for Val=1:length(V1)
   if V1 (Val)==round (V1 (Val)/2)*2
      VE=VE+1;
   end
end
```

(d) What is stored in VC? HINT: Look up the sign function.

```
VD=0;
V1=[9 -4 -3 3 6 -1 0 5 5 -2 -2 -3];
for Val=2:length (V1)
   VD(Val)=1;
   if sign(V1(Val)) ~=sign(V1(Val-1))
      VD (Val)=-1;
   end
   if abs(V1 (Val))==abs(V1 (Val-1))
      VD(Val)=0;
   end
end
```

ICA 20-4

You wish to design a program to tabulate values for the function $V = e^{-t/\tau} \cos \theta t$, a common type of voltage response in electric circuits.

(a) The user should be asked to enter the following parameters:

- τ – time constant in seconds
- θ – frequency in radians per second
- t_0 – initial time in seconds
- t_f – final time in seconds
- *Steps* – number of divisions into which the range of time is to be divided.

(b) The program should print a statement similar to the following:

```
Calculations of the equation e^(-t/x.xxx) cos (x.xxx t) from
x.xxx to x.xxx seconds.
```

(c) The program should calculate the value of V [voltage] for the stated values of τ and θ over the range of time from t_0 to t_f. The results should be printed in a neatly formatted table showing both time and voltage.

ICA 20-5

Assume the vector `AM` contains an even number of elements. Write a short section of code that will divide the product of the even elements by the sum of the odd elements and place the result in `PDS`. Note that this must work correctly for any vector as long as it contains an even number of elements. Write the code using.

(a) a `for` loop without using built-in functions like `sum` or `prod`

(b) direct matrix operations, no loops.

EXAMPLE: `AM =[9 3 5 0.5 10 4]`
`PDS = 0.25` Detailed calculation: (3*0.5*4)/(9+5+10) = 0.25

ICA 20-6

Write a function named `CLASS` that will accept a vector of unknown length called `INT` and return four variables containing the following information about the contents of the vector:

- `NotInt` – contains the number of values in `INT` that are not integers.
- `NotPos` – contains the number of values in `INT` that are zero or less.
- `TooBig` – contains the number of values in `INT` that are greater than 99. Note that `NotInt` has priority over `NotPos` and `TooBig`. For example, an element in `INT` equal to -9.43 would add one to `NotInt`, but NOT to `NotPos`.
- `Ints` – a 99 element vector in which each element contains the number of occurrences of the element's index value in `INT`. For example, if `INT` contains a total of fourteen 57s, `Ints`(57) will contain 14.

Example:

```
INT=[-3, -7.3, 0, 16, 2, 99, 11.3, 298, 176.98, 16]
  NotInt = 3
  NotPos = 2
  TooBig = 1
  Ints(2) = 1
  Ints(16) = 2
  Ints(99) = 1
```

All other elements of `Ints` contain a zero.

Note that the sum of all elements in the returned variables should equal the number of elements in INT.

You MAY NOT use any built in functions to count the number of occurrences of each integer between 1 and 99, nor may you use any function whose sole purpose is to determine if a value is an integer or not. You MAY use other functions, however.

ICA 20-7

For each code segment below, determine the contents of the specified variable following execution. If an error will occur, write ERROR and explain what caused the error.

(a) What is stored in A?

```
for i=1:3
   for j=1:4
      A(j,i)=i^2-j;
   end
end
```

(b) What is stored in SP?

```
M=[1 2 3;4 5 6];
[Rows,Cols]=size(M);
SP(1)=0;
for r=1:Rows
    P=1;
    for c=1:Cols
        P=P*M(r,c);
    end
    SP(r+1)=SP(r)+P;
end
```

(c) What is stored in PR?

```
M=[0 2 4 6;1 -1  2 -2; -3  0 -2  0];
[R,C]=size(M);
PR=zeros(R,C);
for row=1:1:R
   for col=1:1:C
      PR(row, col)=(PR(row, col)+M(row, col))*max(M(row, :));
   end
end
```

ICA 20-8

Write a function named GetArray that will accept two positive integers R and C representing the number of rows and columns of a matrix NA, and return the R × C matrix NA with values input by the user. For each value to be input a prompt should appear similar to the following:

```
Please input the value for row 2 and column 4:
```

with, of course, the 2 and 4 replaced by the indices of the actual element being entered.

ICA 20-9

Write a function named `EvenSum` that will accept a matrix of unknown size and return the sum of all elements at the intersection of even numbered rows and even numbered columns. Note that your function must work with any size matrix. Also note that if the matrix is a scalar or a vector, the function will return zero since there are no even numbered rows or columns in those cases.

EXAMPLE: For the matrix shown, `A=EvenSum(Test);` will place the value 13 in `A`. The numbers added together are shown in **red**.

$$\text{Test} = \begin{bmatrix} 1 & 3 & 5 & 7 & 9 \\ 0 & 1 & 2 & 3 & 4 \\ 6 & 4 & 2 & 1 & 8 \\ 3 & 5 & 2 & 4 & 0 \end{bmatrix}$$

ICA 20-10

You are assessing the price of various components from different vendors and wish to find the least expensive vendor for each component. The prices of the parts from each vendor are stored in a matrix, `VendCost`. Each row corresponds to a specific vendor and each column corresponds to a specific component. If a specific part is not offered by a vendor, the corresponding entry will be `-1`.

Write a program that will determine which vendor offers the cheapest price for each component, and place the results in a two row matrix `Cheapest` with the same number of columns as there are columns in `VendCost`. Each entry in the first row of `Cheapest` should be an integer corresponding to the row number of the vendor with the cheapest price for the corresponding component and the entries in row 2 should contain the lowest price for that component. You may assume that each part is available from at least one of the listed vendors. If two or more vendors offer a component at the same lowest price, you may choose either one.

You may not use the built-in `min` function or other similar functions to solve this problem. You may not use direct matrix operations to solve this problem; you must do it using `for` loops (in a meaningful way). Your solution must work for any number of vendors and any number of components.

Example:

```
VendCost =
    4.97    8.54    2.04    0.44    13.55    -1.00
    5.23    8.23    2.12    0.39    15.98     2.67
    5.24    8.22    2.09    0.51    -1.00     2.76
Cheapest =
    1       3       1       2        1       2
    4.97    8.22    2.04    0.39    13.55    2.67
```

ICA 20-11

You are studying the number of defective parts produced each week by several machines to help adjust maintenance protocols.

Assume the rows of matrix `Def` represent different machines and all columns except the last represent weeks. The last column contains the long-term average of the number of defects per week produced by that machine. Write a short section of MATLAB code that will generate a new matrix `Comp` with the same number of rows but one fewer columns as described.

The code will compare each value in the matrix, except those in the last column, to the value in the last column of the same row to compare the number of defective parts produced by each machine each week with that machine's long-term defect rate.

- If the number of errors equals that machine's average, the corresponding element in the new matrix `Comp` will equal 0.
- If the number of errors is greater than that machine's average, the corresponding element in the new matrix `Comp` will equal 1.
- If the number of errors is less than that machine's average, the corresponding element in the new matrix `Comp` will equal −1.

ICA 20-12

The Pascal Triangle has an amazing number of uses, from linear algebra to design of fractal antennas. Write a program following the five steps listed that will create a Pascal triangle.

1. Prompt the user to enter an integer and place it in a variable `N`.
2. Create an `N` × `N` matrix named `Pascal` filled with zeros.
3. Modify the `Pascal` matrix so that the first row and first column contain all ones.
4. The value of every other element of the `Pascal` matrix should be changed to the sum of the element to the left and the element above. You should complete row 2, then row 3, etc. until all rows have been calculated. Example for `N` = 5 shown.

NOTE: You MAY NOT use the `pascal` function built in to MATLAB.

```
          1   1   1   1   1
          1   2   3   4   5
Pascal =  1   3   6  10  15
          1   4  10  20  35
          1   5  15  35  70
```

5. Reset all elements of the `Pascal` matrix below the minor diagonal (from lower left to upper right) to zero. Example for `N` = 5 shown.

```
          1   1   1   1   1
          1   2   3   4   0
Pascal =  1   3   6   0   0
          1   4   0   0   0
          1   0   0   0   0
```

ICA 20-13

Write a function named `ProdStats` that will accept a matrix `ProdData` that has at least one row and at least two columns. The rows each represent a different machine in a factory, and the columns represent successive days of production. The value in each element is the number of units produced by that machine on the given day.

`ProdStats` should return a new matrix `Trend` with the same dimensions as `ProdData`, and a row vector `TrendNum` with three elements.

Each element of `Trend` indicates whether that day's production for that machine was less than, equal to, or greater than the previous day's production for that machine. Since the first day (first column) does not have previous data upon which to base the comparison, the first column of `Trend` will arbitrarily be set to all zeros. All other elements will be set to either `-1` (lower production than previous day), `0` (equal production to previous day), or `1` (higher production than previous day).

The first element of TrendNum contains the total number of days that production decreased (in other words, the number of negative ones in Trend), the second element of TrendNum contains the number of days with no change in production (zeros in Trend, NOT counting the zeros on the first day), and the third element contains the number of days with higher production (ones in Trend).

Example:

$$\texttt{ProdData} = \begin{bmatrix} 1 & 2 & 4 & 4 & 5 & 4 \\ 7 & 8 & 9 & 0 & 1 & 5 \\ 9 & 8 & 5 & 6 & 7 & 7 \end{bmatrix}$$

$$\texttt{Trend} = \begin{bmatrix} 0 & 1 & 1 & 0 & 1 & -1 \\ 0 & 1 & 1 & -1 & 1 & 1 \\ 0 & -1 & -1 & 1 & 1 & 0 \end{bmatrix} \qquad \texttt{TrendNum} = [4 \;\; 2 \;\; 9]$$

ICA 20-14

For each code segment below, determine the contents of the specified variable following execution. If an error will occur, write ERROR and explain what caused the error.

(a) What is stored in Q?

```
Q=[];
N=1;
while length(Q)<4
   Q=[Q,N]*2;
   N=N*3;
end
```

(b) What is stored in IT?

```
IT=0;
for  k=3:2:7
     C=k^2;
     while C>=0
          C=C-25;
          IT=IT+1;
     end
     IT=IT*2;
end
```

(c) What is stored in R?

```
R=1;
N=1;
while R>1E-6
     R=1;
     for  K=1:N
          R=R/K;
     end
N=N+1;
end
```

ICA 20-15

Assume that a simple menu has been created by the following line of code:

```
SystemStatus=menu('System Status','ON','OFF');
```

Write a short section of code that will handle the situation if the user closes the menu instead of making a selection. If the menu is closed without a proper selection, the user should be told to try again, then the menu should be redisplayed. This should continue until the user makes a proper selection.

ICA 20-16

Write a program that will ask the user to input a single number `N`. If the number is nonzero, a one hundred element row vector named `SEQ` will be filled with the values 0.01*N, 0.02*N, . . . , 0.99*N, N. Note that this must work for both positive and negative values of `N`. After filling the vector, the program should ask if this is correct, or does the user want to enter a new value for `N`, in which case it should refill `SEQ` with the corrected values. This should continue until the user enters a zero, at which point the program terminates.

ICA 20-17

You are writing the code to control a chemical reaction. Sensors are used to place values into the variables `Temp` (temperature in degrees Celsius), `Pres` (pressure in atmospheres), `pH` (pH of solution), and `Time` (elapsed time in minutes since start of reaction). A function called `Update` reads the sensors and modifies the contents of the variables to indicate the current conditions. To use the `Update` function, you simply include the line

```
[Temp,Pres,pH,Time]=Update;
```

in your code as needed.

Write a program that will use the `Update` function once every minute to read new values from the sensors, and continue doing so as long as the pH has decreased by at least 0.02 since the last reading. (Hint: Remember the `pause` function – `pause(n)` pauses execution for n seconds.)

If at any time the pressure exceeds 5 atmospheres or the temperature exceeds 200 degrees Celsius, a message should be generated indicating which parameter is beyond the safe limits and the parameter value. In this case, the program should call the function `terminate` which will shut down the reaction and the program stops after printing a message to the screen such as

```
Reaction terminated after X minutes.
Final pH is x.xxx.
```

where X and x.xxx are replaced with the elapsed time and the final pH.

If at any point, the pressure exceeds 4 atmospheres or the temperature exceeds 175 degrees Celsius, an appropriate warning should be generated stating which parameter is too high and what its value is, but the reaction should continue and the program keep running.

When the pH has decreased by less than 0.02 from the previous reading, the reaction should be terminated, a message printed giving elapsed time and final pH as above, and the program should end.

You may assume that the initial pH is 8.0, the initial temperature is 30 degrees Celsius the initial pressure is 1 atmospheres, and the initial time is 0.

ICA 20-18

You are studying the effect of experience on the productivity of workers.

For a group of new workers, you record how many units each person completes per day for several weeks. You make the assumption that when the average of three consecutive days is not greater than 2% more than the maximum daily production prior to those three days, then that worker has reached maximum productivity, and you do not need to continue looking at the data for subsequent days.

The provided file, `workers.mat`, contains an array with this productivity data. Each row represents a specific worker, and the columns represent successive days. The value in each element of the matrix represents the number of units produced by a specific worker on a specific day.

Your program should analyze the data for each worker and print the following message for each worker, one per line:

```
Worker # X reached maximum productivity after Y days.
```

where X is the worker number (same as row number) and Y is the number of days where the above mentioned criterion is first reached.

If a worker has not reached maximum productivity after the trial period represented by the data, the following message should be produced instead:

```
Worker # X did not reach maximum productivity within Y days.
```

where in this case Y is the total number of days in the trial period.

Note that although this problem can be solved using `for` loops and `if` statements without a `while` statement, you are expected to use a `while` loop to stop scanning each worker's record when the day of maximum productivity has been found. (You may also use `for` loops and `if` statements as necessary, of course.)

Your code should work for any number of workers and any number of days, not just the sample data provided.

ICA 20-19

While experimenting with coding sequences, you decide to try a modification of the factorial sequence by calculating the product of consecutive odd integers instead of all consecutive integers. You call this sequence `OFact`. The first four values of the `OFact` sequence are:

$$OFact(1) = 1$$
$$OFact(2) = 1 * 3 = 3$$
$$OFact(3) = 1 * 3 * 5 = 15$$
$$OFact(4) = 1 * 3 * 5 * 7 = 105$$

Write a program that will ask the user for a maximum sequence value desired, and then print the `OFact` sequence to the command window as long as the value calculated does not exceed the maximum value entered by the user. For example, if the user specifies a maximum of 100, it would print the first three values since `OFact(4) = 105 > 100`. This should be printed in two columns: `N` in the first column and `OFact(N)` in the second column. All values generated should be stored in the vector `OFact` for later use.

When the program reaches the maximum value, a message should be printed that states:

```
The desired maximum was XXX.
OFact(Y) = ZZZ is the closest to this value without exceeding it.
```

where XXX is the maximum entered by the user, Y is the sequence number of the calculated value, and ZZZ is the last calculated value.

CHAPTER 20 REVIEW QUESTIONS

1. Write a function for finding the maximum value in a vector of data. It should receive an array as an argument and return the maximum of the values. Do not use the `max` built-in function.

2. A matrix named `mach` contains three columns of data concerning the energy output of several machines. The first column contains an ID code for a specific machine, the second column contains the total amount of energy produced by that machine in calories, and the third column contains the amount of time required by that machine to produce the energy listed in column 2 in hours.

   Write a function named `MPower` to accept as input the matrix mach and return a new matrix named `P` containing two columns and the same number of rows as `mach`. The first column should contain the machine ID codes, and the second column should contain the average power generated by each machine in units of watts.

3. The Fibonacci sequence is an integer sequence calculated by adding previous numbers together to calculate the next value. This is represented mathematically by saying that $F_n = F_{n-1} + F_{n-2}$ (where F_n is the nth value in the sequence, F) or:

		0 + 1=	1 + 1=	1 + 2=	2 + 3=	3 + 5=	5 + 8=	8 + 13=	$F_{n-1} + F_{n-2}$=
0	**1**	**1**	**2**	**3**	**5**	**8**	**13**	**21**	**. . . F_n**

Note this sequence starts with the underlined values (0, 1) and calculates the remaining values in the sequence based on the sum of the previous two values.

Professor Bowman found this sequence to be extremely insufficient and created the Bowman sequence, which is an integer sequence calculated by adding the previous three numbers together (instead of two like in the Fibonacci sequence) to calculate the next value. This is represented mathematically by saying that $F_n = F_{n-1} + F_{n-2} + F_{n-3}$ (where F_n is the nth value in the sequence, F) or:

			0 + 1 + 2=	1 + 2 + 3=	2 + 3 + 6=	6 + 11 + 20=	$F_{n-3} + F_{n-2} + F_{n-1}$=
0	**1**	**2**	**3**	**6**	**11**	**37**	**. . . F_n**

Note this sequence starts with the underlined values (0, 1, 2) and calculates the remaining values in the sequence based on the sum of the previous three values.

Write a MATLAB function that implements the Bowman sequence that accepts one input argument, the length of the desired Bowman sequence to generate, and returns one output variable, the Bowman sequence stored inside of an array. This function should also check to see if the number passed in to the function is a valid Bowman sequence length (think about what might constitute valid sequence lengths!). If the input is invalid, your function should display an error message and the output variable should contain only one number: –1. Otherwise if the input is valid, your function should calculate the Bowman sequence and display each value in the sequence in the Command Window.

Sample Output:

```
>> B=FUNCTIONNAME(8);
      Bowman sequence:
      0    1    2    3    6    11    20    37
```

4. You are working for a data analytics firm that has been asked to create a generic data collection tool that will provide basic statistics on input typed in to the screen. This data collection tool should ask the user to type the number of desired data points they need to record, the number of decimal places for all of the numbers displayed in the final output, then the program should allow them to record all of the data points into the Command Window, one by one, where the user presses the Enter key after each data point. You may assume that this program will only need to handle numeric values and will not need to worry about strings, vectors, or input of other variable types.

As soon as the user inputs all of the values, your program should generate a string representation of the vector of data, as well as provide basic statistics including the number of negative values, number of positive values, the sum of all of the values, the mean, median, and standard deviation of the data set, and the minimum and maximum values in the data set. The output should appear similar to the output shown below. Note that the vector of data does not need to be displayed with the number of decimal places specified by the user.

Sample Input/Output:

```
Type the number of data points to record: 4
Type the number of decimal places to show in output: 2
Data Point #1: 8
Data Point #2: 2
Data Point #3: 6
Data Point #4: 4

Data Set Information:

Vector = [8 2 6 4]

# Negative: 0.00
# Positive: 4.00

Minimum: 2.00
Maximum: 8.00

Sum: 20.00
Mean: 5.00
Median: 5.00
Standard Deviation: 2.58
```

5. You are part of an engineering firm on contract by the U.S. Department of Energy's Energy Efficiency and Renewable Energy task force to develop a program to help laboratory technicians measure the efficiency of their lab equipment. Your job is to write a program that measures the efficiency of hot plates.

The program will begin by suggesting four possible fluids for the technician to choose from using a menu: acetic acid, citric acid, glycerol, and olive oil. The technician is then prompted to enter the initial room temperature in units of degrees Fahrenheit, the brand name and model of the hot plate, and the theoretical power for the hot plate provided by the manufacturer.

The program will then call a function, which will determine the following:

- All fluid properties [Specific Gravity, Specific Heat] should be contained in your function, not in the main program.
- The technician will use 2 liters of fluid.
- The technician should take 2 data points, one at 2 minutes and one at 5 minutes during the heating process. The technician will then begin to heat the fluid. The program will

prompt the technician to enter each data point at the time interval. The technician will record the temperature of the fluid [degrees Fahrenheit] for the two data measurements.

- Once the final data point has been entered, the function will calculate the energy required to heat the fluid, in joules, and the power used to heat the fluid, in watts
- The function will return to the program the time interval, the temperature readings, and the power used.
- The program will then calculate the efficiency of the hot plate, in percentage.

The output of your program should look like the output displayed below; where the highlighted blue values are example responses provided by the user (typed or by pressing a button depending on the requirements mentioned above) into your program, and the highlighted yellow values are the calculated values that will change based upon your starting properties. Note the DATA SHOWN AS RESULTS AND ON THE GRAPH ARE EXAMPLES ONLY AND MAY NOT REFLECT ACTUAL CALCULATIONS! The code should line up the output calculations. In addition, your code must display the efficiency rounded to the nearest integer and must include a percent symbol.

Finally, the program should produce a graph of the two experimental data points relating time and temperature and a trendline with a formatted equation found using `polyfit`.

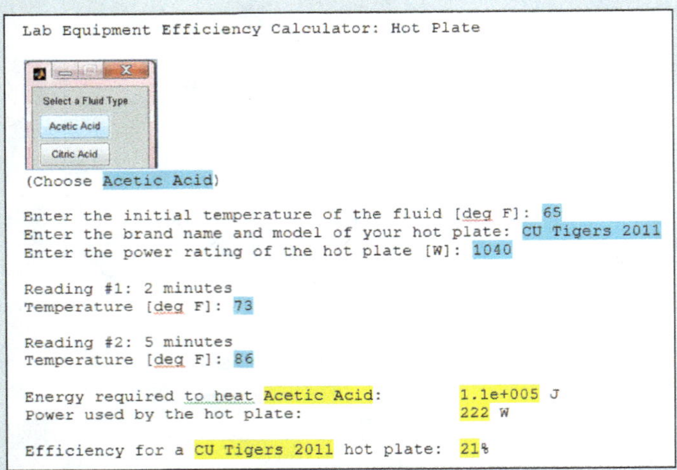

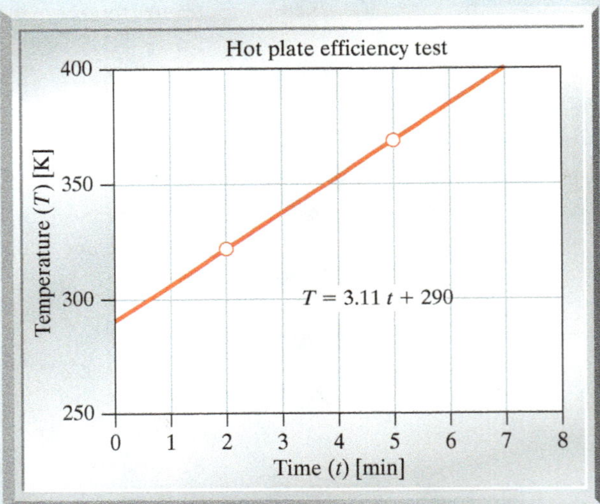

Use the following fluid properties in your program:

Fluid	Specific Gravity [–]	Boiling Point [°C]	Specific Heat [J/(g K)]
Acetic acid	1.049	118	2.18
Citric acid	1.665	153	4
Glycerol	1.261	290	2.4
Olive oil	0.915	300	1.97

6. A sample data set D is provided online in the file `Spacecraft.mat`. Use the `load` command to load this data into your workspace. A matrix D contains data on several experimental spacecraft engines. The matrix is organized in pairs of rows. Each pair of rows contains data measured during tests of the engines. For each pair of rows, the odd numbered row contains several measurements of the total energy used by the engine and the even numbered row contains the corresponding total kinetic energy imparted to the spacecraft. Recalling the equation for energy efficiency, $E_0 = \eta E_I$ we see that the efficiency of the engine is the slope of the line if the kinetic energy of the spacecraft (the output energy, E_0) is plotted versus energy input to the engine(E_I).

For each data set in matrix D:

(a) Determine the efficiency of the engine represented by each data set.
(b) For each data set with an efficiency of at least 0.8 (80%), plot the data (all data sets on the same graph) represented by **red diamonds**. Add a **solid red** trendline to each such data set, including a trendline equation.
(c) For each data set with an efficiency of at least 0.5 (50%) but less than 0.8 (80%), plot the data (all data sets on the same graph) represented by **blue circles**. Add a **dotted blue** trendline to each such data set, including a trendline equation.
(d) For each data set with an efficiency less than 0.5 (50%), plot the data (all data sets on the same graph) represented by **black X's**. These low-efficiency data sets will NOT have a trendline or equation added.
(e) The trendline equations should be in the form $E\#_0 = \eta\ E\#_I$ where # is the number of the dataset (half of the even-numbered row of the data set).
(f) The trendline equations should be positioned immediately to the right of the largest (rightmost) data point of the corresponding data set.

Note that you SHOULD NOT figure out the efficiencies and hard-code the line types, etc. Your program must do this automatically so it will work for any set of data without user intervention.

The graph should be properly labeled and formatted, but the colors and types of both the lines and data markers will not follow the standard proper plot rules since they are specified in the problem.

You may assume the largest value of E_I is 12 and the largest value of E_0 is 10.

For the sample dataset provided, your graph should be similar to the following.

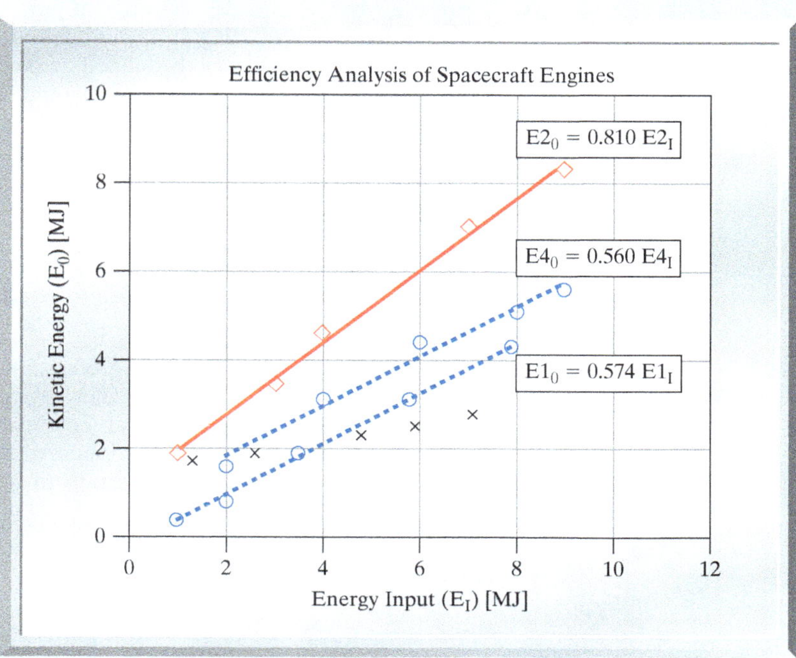

7. You will be given MATLAB file that contains salary data from a major league baseball in 2005. Use the data provided to determine:
   (a) Salary mean, median, and standard deviation for all players.
   (b) Salary mean, median, and standard deviation for the Arizona Diamondbacks.
   (c) Salary mean, median, standard deviation for all pitchers.
   (d) Salary mean, median, standard deviation for all outfielders.
   (e) Which team had the highest average salary?
   (f) Which position had the lowest average salary?
   (g) Draw a histogram and CDF in Excel for all players.
   (h) What percentage of players earned more than $1 million?

8. The following Microsoft Excel file has been provided online. The file contains energy consumption data by energy source per year in the United States, measured in petaBTUs.

Year	Fossil Fuels	Elec. Net Imports	Nuclear	Renewable
2007	101.605	0.106	8.415	6.830
2006	99.861	0.063	8.214	6.922
2005	100.503	0.084	8.160	6.444
2004	100.351	0.039	8.222	6.261
2003	98.209	0.022	7.959	6.150

   (a) Write the MATLAB code necessary to read the Microsoft Excel file and store each column of data into different variables. Create the following:
   - `Yr`: A vector of all of the years in the worksheet.
   - `FF`: A vector of all of the fossil fuels for each year in `Yr`.
   - `ENI`: A vector of all electric imports for each year in `Yr`.
   - `Nuc`: A vector of all nuclear energy consumption for each year in `Yr`.
   - `Ren`: A vector of all renewable energy consumption for each year in `Yr`.
   - `Hdr`: A cell array of all of the headers in row 1.

   You may not hard-code these variables—they should be imported from the Excel file to receive credit.

   (b) Write the MATLAB code necessary to generate the table below of nuclear energy consumption by year using formatted output in the Command Window. You may assume that you have correctly defined the vector `Nuc` and cell array `Hdr` in part **(a)**. Note the nuclear energy consumption should be displayed to two decimal places.

Nuclear Energy Consumption by Year [petaBTU]

2007	2006	2005	2004	2003	. . . etc
8.42	8.21	8.16	8.22	7.96	

   (c) Add code to ask the user if they wish to select another type of fuel, or terminate the program. If the user wants to select again, then part (b) should repeat.

9. As early as 650 BC, mathematicians had been composing magic squares, a sequence of n numbers arranged in a square such that all rows, columns, and diagonals sum to the same constant. Used in China, India, and Arab countries for centuries, artist Albrecht Dürer's engraving Melencolia I (year: 1514) is considered the first time a magic square appears in European art. Each row, column, and diagonal of Dürer's magic square sums to 34. In addi-

tion, each quadrant, the center four squares, and the corner squares all sum to 34. An example of a "magic square" is displayed below.

16	3	2	13
5	10	11	8
9	6	7	12
4	15	14	1

Write a program to prove a series of numbers is indeed a 4 × 4 magic square. Your program should complete the following steps, in this order:

(a) Ask the user to enter their proposed magic square in a single input statement (e.g., [1 2 3 4; 5 6 7 8; 9 10 11 12; 13 14 15 16]—note this example is a 4 × 4 matrix, but NOT a magic square). You may assume the user will enter whole numbers; they will not enter either decimal values or text.

(b) Check that all values are positive; ** **`for`-loop or nested `for`-loop required in the solution.** If one or more of the values in the matrix are negative or zero, issue a statement to the command window informing the user of the mistake and exit the program. This check should work even if the user does not enter a 4 × 4 matrix; it should work regardless of the size of matrix entered.

(c) Check for an arrangement of 4 × 4. If the matrix is not a 4 × 4, issue a statement to the command window informing the user of the mistake and exit the program.

(d) Determine if the matrix is a form of a magic square. The minimum requirement to be classified as a magic square is each row and column sums to the same value. ** **`for`-loop or nested for-loop required in the solution.** If this criteria is not met, issue a statement to the command window informing the user they have not entered a magic square and exit the program.

(e) Determine the classification of the magic square using the following requirements:

1. If each row and column sums to the same value, the magic square is classified as "semi-magic"; the summation value is called the magic constant. ** **`for`-loop or nested `for`-loop required in the solution.**
2. If, in addition to criterion 1, each diagonal sums to the same value, the magic square is classified as "normal;" ** **`for`-loop or nested `for`-loop required in the solution. The use of built-in functions such as `diag`, `fliplr`, `rot90`, `trace` or similar built-in functions is forbidden.**
3. If, in addition to #1 and #2, the largest value in the magic square is equal to 16, the magic square is classified as "perfect;"

Format your magic square classification similar to the format shown below. You may choose to format your table differently, but each classification should contain a "yes" or "no" next to each magic square category.

`The magic constant for your magic square is 24. The classification for your magic square:`

Semi-magic	Normal	Perfect
yes	yes	yes

A few test cases for you to consider:

- Albrecht Dürer magic square: [16, 3, 2, 13; 5, 10, 11, 8; 9, 6, 7, 12; 4, 15, 14, 1];
- Chautisa Yantra magic square: [7, 12, 1, 14; 2, 13, 8, 11; 16, 3, 10, 5; 9, 6, 15, 4];
- Sangrada Familia church, Barcelona magic square: [1, 14, 14, 4; 11, 7, 6, 9; 8, 10, 10, 5; 13, 2, 3, 15];
- Random magic square: [80, 15, 10, 65; 25, 50, 55, 40; 45, 30, 35, 60; 20, 75, 70, 5];
- Steve Wozniak's magic square: [8, 11, 22, 1; 21, 2, 7, 12; 3, 24, 9, 6; 10, 5, 4, 23].

10. Write a program to prove a series of numbers is indeed a 4 × 4 magic square. Your program should complete the following steps, in this order:

(a) Ask the user to enter their proposed magic square in a single input statement (e.g., [1 2 3 4; 5 6 7 8; 9 10 11 12; 13 14 15 16]—note this example is a 4 × 4 matrix, but NOT a magic square). You may assume the user will enter whole numbers; they will not enter either decimal values or text.

(b) Check that all values are positive; ** **`for`-loop or nested `for`-loop required in the solution.** If one or more of the values in the matrix are negative or zero, issue a statement to the command window informing the user of the mistake and ask the user to enter another matrix. This check should be repeated until the user enters a matrix with positive values. This check should work even if the user does not enter a 4 × 4 matrix; it should work regardless of the size of matrix entered.

(c) Check for an arrangement of 4 × 4. If the matrix is not a 4 × 4, issue a statement to the command window informing the user of the mistake and ask the user to enter another matrix. This check should be repeated until the user enters a 4 × 4 matrix. You may assume the re-entered matrix contains only positive values; you do not need to re-check the new matrix for positive values, only for matrix dimensions.

(d) Determine if the matrix is a form of a magic square. The minimum requirement to be classified as a magic square is each row and column sums to the same value. ** **`for`-loop or nested `for`-loop required in the solution.** If this criteria is not met, issue a statement to the command window informing the user they have not entered a magic square and ask the user if they wish to try another magic square. This question can be posed using either a text answer entered by the user (yes, no) or by using a menu. If the user chooses to run the program again, the entire program starting with step (a) should begin again.

(e) Determine the classification of the magic square using the following requirements:

1. If each row and column sums to the same value, the magic square is classified as "semi-magic"; the summation value is called the magic constant. ** **for-loop or nested for-loop required in the solution**.
2. If, in addition to criterion 1, each diagonal sums to the same value, the magic square is classified as "normal;" ** **`for`-loop or nested `for`-loop required in the solution. The use of built-in functions such as `diag, fliplr, rot90, trace` or similar built-in functions is forbidden.**
3. If, in addition to #1 and #2, the largest value in the magic square is equal to 16, the magic square is classified as "perfect;"

Format your magic square classification similar to the format shown below. You may choose to format your table differently, but each classification should contain a "yes" or "no" next to each magic square category.

```
The magic constant for your magic square is 24. The classification
for your magic square:
```

Semi-magic	Normal	Perfect
yes	yes	yes

After this table appears, ask the user if they wish to try another magic square. This question can be posed using either a text answer entered by the user (yes, no) or by using a menu. If the user chooses to run the program again, the entire program starting with step (a) should begin again.

A few test cases for you to consider:

- Albrecht Dürer magic square: [16, 3, 2, 13; 5, 10, 11, 8; 9, 6, 7, 12; 4, 15, 14, 1];
- Chautisa Yantra magic square: [7, 12, 1, 14; 2, 13, 8, 11; 16, 3, 10, 5; 9, 6, 15, 4];
- Sangrada Familia church, Barcelona magic square: [1, 14, 14, 4; 11, 7, 6, 9; 8, 10, 10, 5; 13, 2, 3, 15];
- Random magic square: [80, 15, 10, 65; 25, 50, 55, 40; 45, 30, 35, 60; 20, 75, 70, 5];
- Steve Wozniak's magic square: [8, 11, 22, 1; 21, 2, 7, 12; 3, 24, 9, 6; 10, 5, 4, 23].

11. A zombie picks up a calculator and starts adding odd whole numbers together, in order: 1 + 3 + 5 + . . . etc. What will be the last number the zombie will add that will make the sum on his calculator greater than 10,000? Your task is to write the MATLAB code necessary to solve this problem for the zombie or he will eat your brain. The user should be asked to enter the target number (10,000), although your code should be written in such a way that if a target number other than 10,000 is entered, the correct answer for the value entered will be determined.

12. Write a function called `Balloon` that will accept a single variable named `S`. The function should replace `S` with the square of `S` and repeat this process until `S` is either greater than 10^{15} or less than 10^{-15}. The function should return a value `Q` containing the number of times `S` was squared during this process.

Examples:

- If $S = 100$, Q = 3 (S equals 10^{16} after the third square).
- If $S = 0.1$, Q = 4 (S equals 10^{-16} after the fourth square).
- If $S = 3$, Q = 5 (S equals 1.853×10^{15} after the fifth square).

13. You have written three functions for three different games. The names of the games (and the functions that implement them) are Dunko, Bouncer, and Munchies. Each function will accept an integer between one and three indicating the level of difficulty (1 = easy, 3 = hard), and when play is complete will return a text string, either "won" or "lost," to the program that executed the function.

Write a program that will use a menu to ask the user if he or she wants to play a game. If not, the program should terminate. If so, the program should generate another menu to allow the user to select one of the three games by name.

After selecting a game, the program should display another menu asking the user for the desired level of difficulty (Easy, Moderate, or Hard), and the game should begin by calling the appropriate function.

When the user has finished playing the selected game, a message should be displayed indicating whether the user won or lost the game. A menu should then be generated asking if the user wishes to repeat the game just played. If so, the difficulty level menu should be displayed again and the game repeated.

If the user does not wish to repeat the same game, the program should display a menu asking if the user wants to play another game. If so, the game selection menu should be generated again, followed by level selection, etc. If the user does not wish to play another game, the program should display a message saying "Thanks for playing" and terminate.

14. You are to program part of the interface for a simple ATM. When the user inserts their card and types the correct PIN (you do NOT have to write this part of the program), the system will place the users' checking account balance in a variable `CBal` and the users' savings account balance in `SBal`.

You are to write a function that will accept `SBal` and `CBal` as inputs and return two variables `NewCBal` and `NewSBal` containing the checking and savings balances after the transaction is completed. The function should do the following:

- Display a menu titled "Main Menu" with the following three options.
  - Get cash
  - Get balance
  - Quit
- If "Get cash" is selected, another menu titled "Withdrawal amount" with the following four items is displayed:
  - $20
  - $60
  - $100
  - $200

- After selecting an amount, a menu titled "From which account?" should be displayed showing the following two options:
  - Checking
  - Savings
- At this point, the program should verify that the selected account contains sufficient funds for the requested withdrawal.
- If not, a message should be displayed that says: "Sorry. You do not have sufficient funds in your SSSS account to withdraw $XX" where SSSS is either Savings or Checking and $XX is the selected withdrawal amount.
- If funds are available, the program should call a function named `Disp20(x)`, where `x` is the number of $20 bills to dispense. (See note about `Disp20` below.) After that, the withdrawal amount should be subtracted from the appropriate balance.
- After processing the "Get cash" request, the program should return to the main menu.

Note About `Disp20(x)`: The purpose of this function is to dispense the requested number of $20 bills—that is, to shove `x` bills out of the slot in the ATM machine. This does not really exist, since we do not have an ATM machine to work with. Thus, if you try to run your code, you will get an error ("Undefined function . . .").

In order to test your program, add the following function to your current path:

```
function[]=Disp20(x)
fprintf('%.0f $20 bills were dispensed.\n',x)
```

where `x` is the number of bills to be dispensed.

This allows you to know if the program reached the proper location in the code. It is fairly common in software development to use a "dummy" function in the place of a real one when the device to be controlled has not been completed or is not available in order to help verify whether the software is reaching the correct places in the code for various situations.

- If "Get Balance" is selected, another menu titled "Which account?" should appear with the two choices:
  - Checking
  - Savings

  and the program should then display "Your SSSS balance is $bb.bb.," where SSSS is either Savings or Checking and $bb.bb is the balance in the selected account.
- After processing the "Get balance" request, the program should return to the main menu.
- If Quit is selected, the function should return to the calling program with the updated balances in `NewCBal` and `NewSBal`. Note that the new balances will be equal to the original balances if no money was drawn from an account, but they must still be returned in the two new balance variables.

15. For this assignment, you will need the Cincinnati Reds player data file CincinnatiBaseball2010.mat. The data in this file is a capture of the 2010 season on Baseball-Reference.com (http://www.baseball-reference.com/teams/CIN/2010.shtml).

In the MATLAB workspace provided, there are three variables of interest:

- `PlayerData`, a 37×9 matrix which contains:
  - Column 1: The age of the baseball player
  - Column 2: Games played or pitched
  - Column 3: At bats
  - Column 4: Number of runs scored/allowed
  - Column 5: Singles hit/allowed
  - Column 6: Doubles hit/allowed
  - Column 7: Triples hit/allowed
  - Column 8: Home runs
  - Column 9: Runs batted in

Note that each row in `PlayerData` represents a different baseball player.

- `PlayerNames`, a 37 × 1 "cell array" that contains the names of each player in the `PlayerData` matrix.
- `PlayerPositions`, a 37 × 1 "cell array" that contains the position abbreviation of each player in the `PlayerData` matrix.

(a) Create a function that will accept a single input, a cell array of player positions for an entire team—with our data set, this would be a variable like `PlayerPositions`. Your function must create a new cell array that contains only the unique positions, sorted alphabetically.

(b) Create a program that will display the average age, at bats, home runs, and runs batted in for user-selected positions. In order to allow the user to select the positions, you must use the menu function to allow the user to click positions to include in the analysis, as well as a "Done" button to allow the code to continue to display the output.

Sample Output: Assuming the user selects all positions

Position Stats for 2010 Cincinnati Reds

Pos	Ave(Age)	Ave(AB)	Ave(HR)	Ave(RBI)
1B	25	288	19	58
2B	29	626	18	59
3B	31	242	8	39
C	32	197	5	32
CF	25	514	22	77
IF	27	3	1	4
LF	29	338	11	52
MI	24	38	1	2
OF	27	123	4	11
P	27	21	0	1
RF	23	509	25	70
SS	31	347	5	34
UT	36	23	2	2

Sample Output: Assuming user presses "3B," "OF," "Done"

Position Stats for 2010 Cincinnati Reds

Pos	Ave(Age)	Ave(AB)	Ave(HR)	Ave(RBI)
3B	31	242	8	39
OF	27	123	4	11

(c) MLB.com is consistently ranked one of the top 500 websites in the United States, bringing in millions of visitors a year. However, recent surveys have shown that MLB.com is one of the slowest top 500 websites and is looking for new solutions to speed up their website. After hearing about the baseball statistics programs we have developed in our class, MLB.com has contracted us to help them create a MATLAB program to help them improve the user experience of their website. Recreate the analysis in part **(b)**, except this time export the output to a Microsoft Excel Workbook.

Sample Output: Assuming the user selects all positions

	A	B	C	D	E	F
1	Position Stats for 2010 Cincinnati Reds					
2	Pos	Ave(Age)	Ave(AB)	Ave(HR)	Ave(RBI)	
3	1B	25	288	19	58	
4	2B	29	626	18	59	
5	3B	31	242	8	39	
6	C	32	197	5	32	
7	CF	25	514	22	77	
8	IF	27	3	1	4	
9	LF	29	338	11	52	
10	MI	24	38	1	2	
11	OF	27	123	4	11	
12	P	27	21	0	1	
13	RF	23	509	25	70	
14	SS	31	347	5	34	
15	UT	36	23	2	2	
16						

Sample Output: Assuming user presses "3B," "OF," "Done"

	A	B	C	D	E	F
1	Position Stats for 2010 Cincinnati Reds					
2	Pos	Ave(Age)	Ave(AB)	Ave(HR)	Ave(RBI)	
3	3B	31	242	8	39	
4	OF	27	123	4	11	
5						

COMPREHENSION CHECK ANSWERS

CHAPTER 15

CC 15-1

Known:

- The minimum value in the sum will be 2
- The maximum value in the sum will be 20
- Only even numbers will be included

Unknown:
- The sum of the sequence of even numbers

Assumptions:

- [None]

CC 15-2

Known:

- The minimum value in the product of powers of 5 will be 5
- The maximum value in the product of powers of 5 will be 50

Unknown:

- The product of the sequence of powers

Assumptions:

- Only include integer values

Concerning the actual operation to be performed, there are two alternate assumptions, since the wording is intentionally unclear:

1. Multiply all values between 5 and 50 that are an integer power of 5. In other words, the product of 5 and 25.
2. Multiply all integer powers of 5 with a power between 5 and 50 inclusive. In other words $5^5 * 5^6 * 5^7 * \ldots * 5^{49} * 5^{50}$.

CC 15-3

Known:

- Lowest power of 5 in product is 5
- Highest power of 5 in product is 50

Unknown:

- Product of all integer powers of 5 between the lowest and highest powers

Assumptions:

- The result will not exceed the maximum representable number in the computer. Note that by the time the sequence gets to $x = 17$, the product is approximately one googol (10100).

Algorithm:

1. Set product = 1
2. Set x = 5
3. If x is less than or equal to 50
   - **(a)** Multiply product by 5x
   - **(b)** Add 1 to x
   - **(c)** Return to Step 3 and ask the question again
4. End the process

CC 15-4

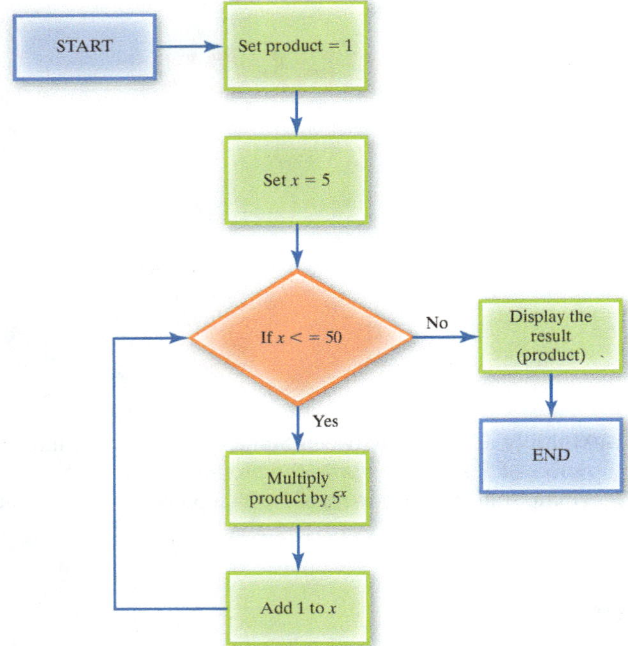

CHAPTER 16

CC 16-1

(a) Valid
(b) Valid
(c) INVALID: cannot begin with number
(d) INVALID: cannot contain special characters (@)
(e) INVALID: cannot contain special characters (-)
(f) Valid

CC 16-2

(a) 2
(b) INVALID: Y not defined; B is on the left side of the =
(c) 57
(d) 3.5
(e) INVALID: Q is not defined
(f) INVALID: Computation not allowed on left of =
(g) INVALID: Computation not allowed on left of =
(h) INVALID: Variable Name is invalid (cannot contain #)
(i) 0.4
(j) 9999
(k) INVALID: Computation not allowed on left of =
(l) 27

CC 16-3

(a) `Int = 8;`
(b) `Real2 = 35.7;`
(c) `Big2 = 47.98E56;`
(d) `Small2 = 3E - 15;`

CC 16-4

(a)

```
% Solution hardcoding coefficients
R1=(-2 + sqrt (2^2 - 4*2*1))/(2*2);
R2=(-2 - sqrt (2^2 - 4*2*1))/(2*2);
```

```
% Better solution using variables
C2=2;
C1=2;
C0=1;
R1=(-C1+sqrt(C1^2 - 4*C2*C0))/(2*C2);
R1=(-C1-sqrt(C1^2 - 4*C2*C0))/(2*C2);
```

(b)

```
% Solution using tand
Trig2=tand(75);

% Solution using tan
Trig2=tan(75*pi/180);

% Solution using degtorad function
Trig2=tan(degtorad(75));
```

CC 16-5

(a) `N4=[17;34;-94;16;0];`
(b) `Tiny=[3.4E-14,9.02E-23,1.32E-9];`
(c) `Ev=2:2:250;`
(d) `Tenths=[10:-0.1:0]';`

CC 16-6

```
Comps=(3*Vals + 5).^4-16;
```

CC 16-7

(a) `TFC=[T1,T2,T3,T4]';`
(b) `Rev=[10:2:10000,700:-7:7];`
(c) `Pow=[RV5,sqrt(RV5),RV5.^2];`

CC 16-8

(a) `Biggie(10:10:50000) = -9999;`
(b)

```
% Solution 1
LV=LV(1:2:length(LV));

% Solution 2
LV(2:2:length(LV))=[];
```

(c) `DS=D(1:2:length(D)) + D(2:2:length(D));`

CC 16-9

(a) `CCM1=[18 0.3;-4.1 -1;0 17];`
(b) `CCM2(2:3,3)=1E15;`
(c) `Corners=CCM3(1:2:3,1:2:3);`

CC 16-10

(a) `CC1=sqrt(QPD/2)`
(b) `CC2=17 + X1 + X2`

CC 16-11

(a) `MTS 'My hero''s hat'`
(b) `B=[blanks(17),'Dr. Willy',blanks(691)]`
(c)

```
% Two different solutions
% Two line solution
Avian=strrep (Avian,'wet','crepuscular');
Avian=strrep (Avian,'never','seldom');
% One line solution
Avian=strrep (strrep(Avian,'never',
'seldom'),'wet','crepuscular');
```

(d)

```
UL=length(Ultimate);
UltAns = str2num(Ultimate(UL-1:UL));
```

CC 16-12

(a) `Cabinet={77,'Dr.Caligari',ones(25,100)}`
(b) `Cabinet={'Diameter','Length';1.5,1:25}`

CC 16-13

(a) `CA{2,1}=CA{1,1}(7:length(CA{1,1}));`
(b) `CA{2,2}=CA{1,2}*CA{1,3};`
(c) `CA{2,3}=CA{1,2}.*CA{1,3}';`

CC 16-14

(a) `BigCell=num2cell(BigMat);`
(b) `VCellMax=cellfun(@max,VCell);`

CC 16-15

```
Resistors(1).Value=100E3;
Resistors(1).Power=1/4;
Resistors(1).Composition='Metal Film';
Resistors(1).Tolerance=0.1;

Resistors(2).Value=2.2E6;
Resistors(2).Power=1/2;
Resistors(2).Composition='Carbon'
Resistors(2).Tolerance=5;

Resistors(3).Value=15
Resistors(3).Power=50;
Resistors(3).Composition='WireWound'
Resistors(3).Tolerance=10;
```

CC 16-16

(a) `MZn=MetalData(1).SpacGrav*1000`
(b) `LtIso=min(MetalData(MNum).Isotopes)`
(c) `save('PTData.mat','Pr1','Pr2','Tmp3')`
(d) `save('a1Var.mat','*a1'))`
(e) `load('VelData.mat','*XS*')`

CC 16-17

(a) `save('TempData.mat')`
(b) `load('PressData.mat')`

CHAPTER 17

CC 17-1

c, d, e, g

CC 17-2

314.1953
`Dogs=areaCircle(10);`

CC 17-3

```
function[R]=RAC (N, A)
  R=3*sin(N^A)
end
```

CHAPTER 18

CC 18-1

```
Height=input('Please enter your height in inches. ');
```

```
TempF=input('Please enter the temperature in degrees Fahrenheit. ');
```

CC 18-2

```
EyeColor=input('Please enter the color of your eyes. ','s');
```

```
Month=input('What is the current month. ','s');
```

CC 18-3

```
Month=menu('The current month is:', 'Jan',
'Feb', 'Mar', 'Apr', 'May', 'Jun', 'Jul',
'Aug', 'Sep', 'Oct', 'Nov', 'Dec');
(all on same line)
C=5
```

CC 18-4

(a) `%0.2f`
(b) `%0.3e or %0.3E`
(c) `0.35`
(d) `3.54e-01 or 3.54E-01`
(e) `000000.354`

CC 18-5

```
Clemson-USC Football Rivalry:

Year      USC     Clemson
2012      27        17
2011      34        13
2010      29         7
```

CHAPTER 19

CC 19-1

(a) `5 <=t && t < 10`
(b) `-30 < M && M <=20`
(c) `Y ~=100`

CC 19-2

(a) $R1 = \begin{bmatrix} 1 & 0 \\ 1 & 1 \\ 1 & 0 \end{bmatrix}$

(b) $R1 = \begin{bmatrix} 0 & -5 \\ 0 & 0 \\ 0 & 0 \end{bmatrix}$

(c) $R1 = \begin{bmatrix} 1 & 1 \\ 1 & 0 \\ 1 & 0 \end{bmatrix}$

CC 19-3

Possible Solution #1:

```
R4=(M1>M2).*M1 + (M2>M1).*M2;
```

Possible Solution #2:

```
LM=M1>M2;
R4=LM.*M1 + ~LM.*M2;
```

CC 19-4

```
if TW <=0
  phase ='solid';
elseif TW >=100
  phase ='gas';
else
  phase ='liquid';
end
```

CC 19-5

```
function[x,y]=SumItUp(a,b,c)
if a + b > 100
  fprintf('The inputs are %0.2f, %0.2f, and
  %0.2f',a,b,c)
else
  fprintf('The inputs are %0.0f, %0.0f, and
  %0.0f', a,b,c)
end
x=a + b;
y=b + c;
```

CC 19-6

Choices (a) and (c).

CC 19-7

```
Phase=menu ('Choose type of water sample',
'Solid','Liquid','Gas');
switch Phase
  case 1
       fprintf('For a solid, water must be
       at a temperature less than 0 degrees
       Celsius.')
  case 3
       fprintf('For a gas, water must be at
       a temperature more than 100 degrees
       Celsius.')
  otherwise
       fprintf('For a liquid, water must be
       at a temperature between 0 and 100
       degrees Celsius.')
end
```

CHAPTER 20

CC 20-1

```
for K=2:2:20
        disp(K)
end
```

CC 20-2

```
for K=5:5:50
        disp(K)
end
```

CC 20-3

```
for K=13:-2:-11
        disp(K)
end
```

CC 20-4

```
PosSum=0;
for E=1:length(Vals)
   if Vals(E)>0
     PosSum=PosSum+Vals(E);
   end
end
```

CC 20-5

```
D=0; % Setup for calculating D
I=1; % Setup for calculating I
% Print column headers
fprintf('k\tA\tB\tC\tD\tE\tF\tG\tH\t\tI\
tJ\n')
for k=1:10
     A=2*k;
     B=k*2-1;
     C=k^2;
     D=D+k; % Note that D was initialized to
          0 before the loop
     E=7+0.25*k;
     F=11-k;
     G=(-1)^(k+1); % There are several other
          ways to do this one
     H=2^k;
     I=I*k; % Note that I was initialized to
          1 before the loop
     J=1/k;
     % Print current row of values
     fprintf('%d\t%d\t%d\t%d\t%d\t%.2f\t%d\
          t%d\t%d\t%9.0f\t%.2f\n',...
          k,A,B,C,D,E,F,G,H,I,J)
end
```

CC 20-6

```
[M1Rows,M1Cols]=size(M1);
PosNums=zeros(1,M1Cols); %initialize
counter vector
for R=1:M1Rows
    for C=1:M1Cols
      if M1(R,C)>=0
        PosNums(C)=PosNums(C)+1
      end
    end
end
```

CC 20-7

```
[M1Rows,M1Cols]=size(M1);
NegRows=M1Rows; %Assume all rows are all
negative
for R=1:M1Rows
    for C=1:M1Cols
      if M1(R,C)>=0
        NegRows=NegRows-1; %Positive value
        found, reduce count
        break  %Go to next row
      end
    end
end
```

CC 20-8

```
% Solution 1
InVal=-1; % Initialize so loop runs first
time
while InVal<0
       InVal=input('Please enter a non-
       negative number: ');
end
```

```
% Solution 2
% This version allows the program to warn
the user they made an error.
InVal=input('Please enter a non-negative
number: '); % First attempt while InVal<0
       disp('You entered an incorrect
       value. Try again.')
       InVal=input('Please enter a non-
       negative number: '); % Next attempt
end
```

CC 20-9

```
Prod=1; % Initialize Prod
Elem=1; % Initialize pointer to vector
elements
while Prod<=1E6 && Elem<=length(V2)
       Prod=Prod*V2(Elem);
       Elem=Elem+1;  %Step to next element
end
```